TEXTBOOK OF GEOCHEMISTRY

TEXTBOOK OF GEOCHEMISTRY

By
Dr. Shardendu Kislaya

DPH

DISCOVERY PUBLISHING HOUSE PVT. LTD.
NEW DELHI-110 002

Published by:
Tilak Wasan
DISCOVERY PUBLISHING HOUSE PVT. LTD.
4383/4B, Ansari Road, Darya Ganj
New Delhi-110 002 (India)
Phone : +91-11-23279245; 23253475; 43596065
E-mail : discoverybooksindia@gmail.com
discoverypublishinghouse@gmail.com
namitwasan9@gmail.com
web : www.discoverypublishinggroup.com

***First Edition:* 2011**
ISBN: 978-81-8356-839-5

Textbook of Geochemistry

Printed at:
Infinity Imaging Systems
Delhi

Preface

The field of geochemistry involves study of the chemical composition of the Earth and other planets, chemical processes and reactions that govern the composition of rocks and soils, and the cycles of matter and energy that transport the Earth's chemical components in time and space, and their interaction with the hydrosphere and the atmosphere.

The most important fields of geochemistry are:

1. *Isotope geochemistry:* Determination of the relative and absolute concentrations of the elements and their isotopes in the earth and on earth's surface.
2. Examination of the distribution and movements of elements in different parts of the earth (crust, mantle, hydrosphere etc.) and in minerals with the goal to determine the underlying system of distribution and movement.
3. *Cosmochemistry:* Analysis of the distribution of elements and their isotopes in the cosmos.
4. *Biogeochemistry*: Field of study focusing on the effect of life on the chemistry of the earth.

5. *Organic geochemistry:* A study of the role of processes and compounds that are derived from living or once-living organisms.

6. Regional, environmental and exploration geochemistry: Applications to environmental, hydrological and mineral exploration studies.

The book contains 13 chapters or various aspects of Geochemistry.

Author

Contents

1 Geochemistry

An Introduction

The field of geochemistry involves study of the chemical composition of the Earth and other planets, chemical processes and reactions that govern the composition of rocks and soils, and the cycles of matter and energy that transport the Earth's chemical components in time and space, and their interaction with the hydrosphere and the atmosphere.

The most important fields of geochemistry are:

- *Isotope geochemistry:* Determination of the relative and absolute concentrations of the elements and their isotopes in the earth and on earth's surface.
- Examination of the distribution and movements of elements in different parts of the earth (crust, mantle, hydrosphere etc.) and in minerals with the goal to determine the underlying system of distribution and movement.
- *Cosmochemistry:* Analysis of the distribution of elements and their isotopes in the cosmos.
- *Biogeochemistry:* Field of study focusing on the effect of life on the chemistry of the earth.

- *Organic geochemistry:* A study of the role of processes and compounds that are derived from living or once-living organisms.
- *Regional, environmental and exploration geochemistry:* Applications to environmental, hydrological and mineral exploration studies.

Victor Goldschmidt is considered by most to be the father of modern geochemistry and the ideas of the subject were formed by him in a series of publications from 1922 under the title 'Geochemische Verteilungsgesetze der Elemente'.

Chemical Characteristics

The more common rock constituents are nearly all oxides; chlorine, sulfur and fluorine are the only important exceptions to this and their total amount in any rock is usually much less than 1%. F. W. Clarke has calculated that a little more than 47% of the earth's crust consists of oxygen. It occurs principally in combination as oxides, of which the chief are silica, alumina, iron oxides, lime, magnesia, potash and soda. The silica functions principally as an acid, forming silicates, and all the commonest minerals of igneous rocks are of this nature. From a computation based on 1672 analyses of all kinds of rocks Clarke arrived at the following as the average percentage composition: SiO_2 = 59.71, Al_2O_3 = 15.41, Fe_2O_3 = 2.63, FeO = 3.52, MgO = 4.36, CaO = 4.90, Na_2O = 3.55, K_2O = 2.80, H_2O = 1.52, TiO_2 = 0.60, P_2O_5 = 0.22, total 99.22%). All the other constituents occur only in very small quantities, usually much less than 1 per cent..

These oxides combine in a haphazard way. The potash and soda, for example, combine to produce feldspars. In some cases they may take other forms, such as nepheline, leucite and muscovite, but in the great majority of instances they are found as felspar. The phosphoric acid with lime forms apatite. The titanium dioxide with ferrous oxide gives rise to ilmenite. Part of the lime forms lime felspar. Magnesia and iron oxides with silica crystallize as olivine or enstatite, or with alumina and lime form the complex ferro-magnesian silicates of which the pyroxenes, amphiboles and biotites are the chief. Any excess of silica above what is required to neutralize the bases will separate out as quartz; excess of alumina crystallizes as corundum. These must be regarded only as general tendencies. It is possible by inspection of a rock analysis to say approximately what minerals the rock will contain, but there are numerous exceptions to any rule which can be laid down.

Mineral Constitution

Hence, we may say that except in acid or siliceous rocks containing 66% of silica and over, quartz will not be abundant. In basic rocks (containing 60% of silica or less) it is rare and accidental. If magnesia and iron be above the average while silica is low olivine may be expected; where silica is present in greater quantity over ferro-magnesian minerals, such as augite, hornblende, enstatite or biotite, occur rather than olivine. Unless potash is high and silica relatively low leucite will not be present, for leucite does not occur with free quartz. Nepheline, likewise, is usually found in rocks with much soda and comparatively little silica. With high alcalis soda-bearing pyroxenes and amphiboles may be present. The lower the percentage of silica and the alkalis the greater is the prevalence of t lime felspar as contracted with soda or potash felspar. Clarke has calculated the relative abundance of the principal rock-forming minerals with the following results: Apatite = 0.6, titanium minerals = 1.5, quartz = 12.0, felspars = 59.5, biotite = 3.8, hornblende and pyroxene = 16.8, total = 94.2%. This, however, can only be a rough approximation. The other determining factor, namely the physical conditions attending consolidation, plays on the whole a smaller part, yet is by no means negligible, as a few instances will prove. There are certain minerals which are practically confined to deep-seated intrusive rocks, e.g. microcline, muscovite, diallage. Leucite is very rare in plutonic masses; many minerals have special peculiarities in microscopic character according to whether they crystallized in depth or near the surface, e.g. hypersthene, orthoclase, quartz. There are some curious instances of rocks having the same chemical composition but consisting of entirely different minerals, e.g. the hornblendite of Gran, in Norway, containing only hornblende, has the same composition as some of the camptonites of the same locality which contain felspar and hornblende of a different variety. In this connection we may repeat what has been said above about the corrosion of porphyritic minerals in igneous rocks. In rhyolites and trachytes early crystals of hornblende and biotite may be found in great numbers partially converted into augite and magnetite. The hornblende and biotite were stable under the pressures and other conditions which obtained below the surface, but unstable at higher levels. In the ground-mass of these rocks augite is almost universally present. But the plutonic representatives of the same magma, granite and syenite contain biotite and hornblende far more commonly than augite.

Acid, Intermediate and Basic Igneous Rocks

Those rocks which contain most silica and on crystallizing yield free quartz are erected into a group generally designated the "acid" rocks. Those again which contain least silica and most magnesia and iron, so that quartz is absent while olivine is usually abundant, form the "basic" group. The "intermediate" rocks include those which are characterized by the general absence of both quartz and olivine. An important subdivision of these contains a very high percentage of alkalis, especially soda, and consequently has minerals such as nepheline and leucite not common in other rocks. It is often separated from the others as the "alkali" or "soda" rocks, and there is a corresponding series of basic rocks. Lastly a small sub-group rich in olivine and without felspar has been called the "ultrabasic" rocks. They have very low percentages of silica but much iron and magnesia.

Except these last practically all rocks contain felspars or felspathoid minerals. In the acid rocks the common felspars are orthoclase, which perthite, microcline, oligoclase, all having much silica and alkalis. In the basic rocks labradorite, anorthite and bytownite prevail, being rich in lime and poor in silica, potash and soda. Augite is the commonest ferro-magnesian of the basic rocks, but biotite and hornblende are on the whole more frequent in the acid.

The rocks which contain leucite or nepheline, either partly or wholly replacing felspar are not included in this table. They are essentially of intermediate or of basic character. We might in consequence regard them as varieties of syenite, diorite, gabbro, etc., in which felspathoid minerals occur, and indeed there are many transitions between syenites of ordinary type and nepheline — or leucite — syenite, and between gabbro or dolerite and theralite or essexite. But as many minerals develop in these "alcali" rocks which are uncommon elsewhere, it is convenient in a purely formal classification like that which is outlined here to treat the whole assemblage as a distinct series.

This classification is based essentially on the mineralogical constitution of the igneous rocks. Any chemical distinctions between the different groups, though implied, are relegated to a subordinate position. It is admittedly artificial by it has grown up with the grown of the science and is still adopted as the basis on which more minute subdivisions are erected. The subdivisions are by no means of equal value. The syenites, for example, and the peridotites, are far less important than the granites, diorites and gabbros. Moreover, the effusive andesites do not always correspond to the plutonic diorites but partly also to the gabbros. As the different kinds of rock, regarded as aggregates of minerals, pass gradually into one another, transitional types are very common and are often so important as to receive special names. The quartz-syenites and nordmarkites may be interposed between granite and syenite, the tonalites and adamellites between granite and diorite, the monzoaites between syenite and diorite, norites and hyperites between diorite and gabbro and so on.

2 Isotope Geochemistry

Isotope geochemistry is an aspect of geology based upon study of the relative and absolute concentrations of the elements and their isotopes in the Earth. Variations in the abundance of these isotopes, typically measured with an isotope ratio mass spectrometer or an accelerator mass spectrometer, can reveal information about the age of a rock or the source of air or water. Isotope ratios can even shed light on chemical processes in the atmosphere. Broadly, the field of isotope geochemistry is divided into two branches: *stable* and *radiogenic* isotope geochemistry.

Stable Isotope Geochemistry

For most stable isotopes, the magnitude of fractionation from kinetic and equilibrium fractionation is very small; for this reason, enrichments are typically reported in "per mil" (‰, parts per thousand). These enrichments (δ) represent the ratio of heavy isotope to light isotope in the sample over the ratio of a standard. That is,

$$\delta^{13}C = \left(\frac{\left(\frac{^{13}\mathrm{C}}{^{12}\mathrm{C}} \right)_{Sample}}{\left(\frac{^{13}\mathrm{C}}{^{12}\mathrm{C}} \right)_{Standard}} - 1 \right) * 1000‰$$

Carbon

Carbon has two stable isotopes, ^{12}C and ^{13}C, and one radioactive isotope, ^{14}C. Carbon isotope ratios are measured against Vienna Pee Dee Belemnite (VPDB) and have been used as an indicator of paleoclimate as well as to track ocean circulation. During photosynthesis, organisms using the C_3 pathway show different enrichments compared to those using the C_4 pathway, allowing scientists not only to distinguish organic matter from abiotic carbon, but also what type of photosynthetic pathway the organic matter was using.

Nitrogen

Nitrogen has two stable isotopes, ^{14}N, and ^{15}N. The ratio between these is measured relative to nitrogen in ambient air. Nitrogen ratios are frequently linked to agricultural activities. Nitrogen isotope data has also been used to measure the amount of exchange of air between the stratosphere and troposphere using data from the greenhouse gas N_2O.

Oxygen

Oxygen has three stable isotopes, ^{16}O, ^{17}O, and ^{18}O. Oxygen ratios are measured relative to Vienna Standard Mean Ocean Water (VSMOW) or Vienna Pee Dee Belemnite (VPDB) Variations in oxygen isotope ratios are used to track both water movement, paleoclimate and atmospheric gases such as ozone and carbon dioxide Typically, the VPDB oxygen reference is used for paleoclimate, while VSMOW is used for most other applications Oxygen isotopes appear in anomalous ratios in atmospheric ozone, resulting from mass-independent fractionation Isotope ratios in fossilized foraminifera have been used to deduce the temperature of ancient seas.

Sulfur

Sulfur has four stable isotopes, with the following abundances: ^{32}S (0.9502), ^{33}S (0.0075), ^{34}S (0.0421) and ^{36}S (0.0002). These abundances are compared to those found in Cañon Diablo troilite. Variations in sulfur isotope ratios are used to study the origin of sulfur in an orebody and the temperature of formation of sulfur–bearing minerals.

Radiogenic Isotope Geochemistry

Lead-lead Isotope Geochemistry

Lead has four stable isotopes - ^{204}Pb, ^{206}Pb, ^{207}Pb, ^{208}Pb and one common radioactive isotope ^{202}Pb with a half-life of ~53,000 years.

Lead is created in the Earth via decay of transuranic elements, primarily uranium and thorium.

Lead isotope geochemistry is useful for providing isotopic dates on a variety of materials. Because the lead isotopes are created by decay of different transuranic elements, the ratios of the four lead isotopes to one another can be very useful in tracking the source of melts in igneous rocks, the source of sediments and even the origin of people via isotopic fingerprinting of their teeth, skin and bones.

It has been used to date ice cores from the Arctic shelf, and provides information on the source of atmospheric lead pollution.

Lead-lead isotopes has been successfully used in forensic science to fingerprint bullets, because each batch of ammunition has its own peculiar $^{204}Pb/^{206}Pb$ vs $^{207}Pb/^{208}Pb$ ratio.

Samarium-neodymium

Samarium-neodymium is an isotope system which can be utilised to provide a date as well as isotopic fingerprints of geological materials, and various other materials including archaeological finds (pots, ceramics).

^{147}Sm decays to produce ^{143}Nd with a half life of 1.06 × 1011 years.

Dating is achieved usually by trying to produce an isochron of several minerals within a rock specimen. The initial ^{143}Nd/^{144}Nd ratio is determined.

This initial ratio is modelled relative to CHUR - the Chondritic Uniform Reservoir - which is an approximation of the chondritic material which formed the solar system. CHUR was determined by analysing chondrite and achondrite meteorites.

The difference in the ratio of the sample relative to CHUR can give information on a model age of extraction from the mantle (for which an assumed evolution has been calculated relative to CHUR) and to whether this was extracted from a granitic source (depleted in radiogenic Nd), the mantle, or an enriched source.

Rhenium-osmium

Rhenium and osmium are chalcophile elements which are present at very low abundances in the crust. Rhenium undergoes radioactive decay to produce osmium. The ratio of non-radiogenic osmium to radiogenic osmium throughout time varies.

Rhenium prefers to enter sulfides more readily than osmium. Hence, during melting of the mantle, rhenium is stripped out, and prevents the osmium-osmium ratio from changing appreciably. This *locks in* an initial osmium ratio of the sample at the time of the melting event. Osmium-osmium initial ratios are used to determine the source characteristic and age of mantle melting events.

Protactinium: Thorium Ratios - ^{231}Pa/^{230}Th

Uranium is well mixed in the ocean, and its decay produces ^{231}Pa and ^{230}Th at a constant activity ratio (0.093).

The decay products are rapidly removed by adsorption on settling particles, but not at equal rates. ^{231}Pa has a residence equivalent to the residence time of deep water in the Atlantic basin (around 1000 yrs) but ^{230}Th is removed more rapidly (centuries). Thermohaline circulation effectively exports ^{231}Pa from the Atlantic into the Southern Ocean, while most of the ^{230}Th remains in Atlantic sediments. As a result, there is a relationship between $^{231}Pa/^{230}Th$ in Atlantic sediments and the rate of overturning: faster overturning produces lower sediment $^{231}Pa/^{230}Th$ ratio, while slower overturning increases this ratio. The combination of d13C and $^{231}Pa/^{230}Th$ can therefore provide a more complete insight into past circulation changes.

Noble Gas Isotopes

Helium-3

Helium-3 was trapped in the planet when it was created. Some 3_{He} is being added by meteoric dust, primarily collecting on the bottom of oceans (although due to subduction, all oceanic tectonic plates are younger than continental plates). However, ^{3}He will be degassed from oceanic sediment during subduction, so cosmogenic ^{3}He is not affecting the concentration or noble gas ratios of the mantle.

Helium-3 is created by cosmic ray bombardment, and by lithium spallation reactions which generally occur in the crust. Lithium spallation is the process by which a high-energy neutron bombards a lithium atom, creating a ^{3}He and a ^{4}He ion. This requires significant lithium to adversely affect the $^{3}He/^{4}He$ ratio.

All degassed helium is lost to space eventually, due to the average speed of helium exceeding the escape velocity for the Earth. Thus, it is assumed the helium content and ratios of Earth's atmosphere have remained essentially stable.

It has been observed that ^{3}He is present in volcano emissions and oceanic ridge samples. How ^{3}He is stored in the planet is under investigation, but it is associated with the mantle and is used as a marker of material of deep origin.

Due to similarities in helium and carbon in magma chemistry, outgassing of helium requires the loss of volatile components (water, carbon dioxide) from the mantle, which happens at depths of less than 60 km. However, ^{3}He is transported to the surface primarily trapped in the crystal lattice of minerals within fluid inclusions.

Helium-4 is created by radiogenic production (by decay of uranium/thorium-series elements). The continental crust has become enriched with those elements relative to the mantle and thus more He4 is produced in the crust than in the mantle.

The ratio (**R**) of ^{3}He to ^{4}He is often used to represent 3He content. **R** usually is given as a multiple of the present atmospheric ratio (**Ra**).

Common values for **R/Ra**:

- Old continental crust: less than 1
- mid-ocean ridge basalt (MORB): 7 to 9
- Spreading ridge rocks: 9.1 plus or minus 3.6
- Hotspot rocks: 5 to 42
- Ocean and terrestrial water: 1
- Sedimentary formation water: less than 1
- Thermal spring water: 3 to 11

$^{3}He/^{4}He$ isotope chemistry is being used to date groundwaters, estimate groundwater flow rates, track water pollution, and provide insights into hydrothermal processes, igneous geology and ore genesis.

- (U-Th)/He dating of apatite as a thermal history tool
- USGS: Helium Discharge at Mammoth Mountain Fumarole (MMF)

Groundwater Isotopes

Tritium/helium-3

Tritium was released to the atmosphere during atmospheric testing of nuclear bombs. Radioactive decay of tritium produces the noble gas helium-3. Comparing the ratio of tritium to helium-3 ($^3H/^3He$) allows estimation of the age of recent ground waters.

- USGS Tritium/Helium-3 Dating
- Hydrologic Isotope Tracers - Helium

CHEMICAL ELEMENT

A chemical element is a type of atom that is distinguished by its atomic number; that is, by the number of protons in its nucleus. The term is also used to refer to a pure chemical substance composed of atoms with the same number of protons. Common examples of elements are iron, copper, silver, gold, hydrogen, carbon, nitrogen, and oxygen. In total, 117 elements have been observed as of 2008, of which 94 occur naturally on Earth. 80 elements have stable isotopes (namely all elements with atomic numbers 1 to 82, except elements 43 and 61 (technetium and promethium). Elements with atomic numbers 83 or higher (bismuth and above) are inherently unstable, and undergo radioactive decay. The elements from atomic number 83 to 94 have no stable nuclei, but are nevertheless found in nature, either surviving as remnants of the primordial stellar nucleosynthesis which produced the elements in the solar system, or else produced as short-lived daughter-isotopes through the natural decay of uranium and thorium.

All chemical matter consists of these elements. New elements of higher atomic number are discovered from time to time, as products of artificial nuclear reactions.

History

Several old philosophies used a set of archetypal classical elements to explain patterns in nature. The term 'element' was originally used to refer to a state of matter (solid/earth, liquid/water, gas/air, and plasma/fire) or a phase of matter (as in the Chinese *Wu Xing*), rather than the chemical elements of modern science. The Greek, Indian (*Tattva* and *Mahabhuta*) and Japanese (*go dai*) traditions essentially had the same five elements: Air, Earth, Fire, Water and Aether.

The term 'elements' (*stoicheia*) was first used by the Greek philosopher Plato in about 360 BCE, in his dialogue Timaeus, which includes a discussion of the composition of inorganic and organic bodies and is a rudimentary treatise on chemistry. Plato assumed that the minute particle of each element corresponded to one of the regular polyhedra: tetrahedron (fire), octahedron (air), icosahedron (water), and cube (earth).

Adding to the four elements of the Greek philosopher Empedocles, in about 350 BC, Aristotle also used the term "element"and conceived of a fifth element called "quintessence", which formed the heavens. Aristotle defined an element as:

> *Element:* one of those bodies into which other bodies can be decomposed and which itself is not capable of being divided into other.

Building on the theory in *circa* 790, Arab/Persian chemist and alchemist, Jabir ibn Hayyan (Geber), postulated that metals were formed out of two elements: sulphur, 'the stone which burns', which characterized the principle of combustibility, and mercury, which contained the idealized principle of metallic properties Shortly thereafter, this evolved into the Arabic concept of the three principles: sulphur giving flammability or combustion, mercury giving volatility and stability, and salt giving solidity.

In the 10th century, Persian physician and alchemist Muhammad ibn Zakariya Razi (Rhazes) wrote the *Doubts concerning Galen,* in which he refuted both the Galenic medical theory of four humours and the underlying ancient concept of four classical elements. He carried out an experiment which would upset these theories by inserting a liquid with a different temperature into a body resulting in an increase or decrease of bodily heat, which resembled the temperature of that particular fluid. Razi noted particularly that a warm drink would heat up the body to a degree much higher than its own natural temperature, thus the drink would trigger a response from the body, rather than transferring only its own warmth or coldness to it. Razi's chemical experiments suggested other qualities of matter, such as "oiliness" and "sulfurousness", or inflammability and salinity, which were not readily explained by the traditional fire, water, earth and air division of elements.

In 1524, Swiss chemist Paracelsus adopted Aristotle's four element theory, but reasoned that they appeared in bodies as Geber's three principles. Paracelsus saw these principles as fundamental, and justified them by recourse to the description of how wood burns in fire. Mercury included the cohesive principle, so that when it left in smoke the wood fell apart. Smoke represented the volatility (the mercury principle), the heat-giving flames represented flammability.

In 1661, Robert Boyle showed that there were more than just four classical elements as the ancients had assumed. The first modern list of chemical elements was given in Antoine Lavoisier's 1789 *Elements of Chemistry*, which contained thirty-three elements, including light and caloric. By 1818, Jöns Jakob Berzelius had determined atomic weights for forty-five of the forty-nine accepted elements. Dmitri Mendeleev had sixty-six elements in his periodic table of 1869.

In 1669, German physician and chemist Johann Becher published his Physica Subterranea. In modification on the ideas of Paracelsus, he argued that the constituents of bodies are air, water, and three types of earth: *terra fluida*, the mercurial element, which contributes fluidity and volatility; *terra lapida*, the solidifying element, which produces fusibility or the binding quality; and *terra pinguis*, the fatty element, which gives material substance its oily and combustible qualities These three earths correspond with Geber's three principles. A piece of wood, for example, according to Becher, is composed of ash and terra pinguis; when the wood is burnt, the terra pinguis is released, leaving the ash. In other words, in combustion the fatty earth burns away.

From Boyle until the early 20th century, an element was defined as a pure substance that cannot be decomposed into any simpler substance Put another way, a chemical element cannot be transformed into other chemical elements by chemical processes. In 1913, Henry Moseley discovered that the physical basis of the atomic number of the atom was its nuclear charge, which eventually led to the current definition. The current definition also avoids some ambiguities due to isotopes and allotropes.

By 1919, there were seventy-two known elements In 1955, element 101 was discovered and named mendelevium in honor of Mendeleev, the first to arrange the elements in a periodic manner. In October 2006, the synthesis of element 118 was reported; however, element 117 has not yet been created in the laboratory.

Description

The lightest elements are hydrogen and helium, both theoretically created by Big Bang nucleosynthesis during the first 20 minutes of the universe in a ratio of around 3:1 by mass (approximately 12:1 by number of atoms). Almost all other elements found in nature, including some further hydrogen and helium created since then, were made by various natural or (at times) artificial methods of nucleosynthesis, including occasionally breakdown activities such as nuclear fission, alpha decay, cluster decay, and cosmic ray spallation.

As of 2006, there are 117 known elements (in this context, "known" means observed well enough, even from just a few decay products, to have been differentiated from any other element). Of these 117 elements, 94 occur naturally on Earth. Six of these occur in extreme trace quantities: technetium, atomic number 43; promethium, number 61; astatine, number 85; francium, number 87; neptunium, number 93; and plutonium, number 94. These 94 elements, and also possibly element 98 californium, have been detected in the universe at large, in the spectra of stars and also supernovae, where short-lived radioactive elements are newly being made.

The remaining 22 elements not found on Earth or in astronomical spectra have been derived artificially. All of the elements that are derived solely through artificial means are radioactive with very short half-lives; if any atoms of these elements were present at the formation of Earth, they are extremely likely to have already decayed, and if present in novae, have been in quantities too small to have been noted. Technetium was the first purportedly non-naturally occurring element to be synthesized, in 1937, although trace amounts of technetium have since been found in nature, and the element may have been discovered naturally in 1925. This pattern of artificial production and later natural discovery has been repeated with several other radioactive naturally occurring trace elements.

Atomic Number

The atomic number of an element, Z, is equal to the number of protons which defines the element. For example, all carbon atoms contain 6 protons in their nucleus; so the atomic number "Z" of carbon is 6. Carbon atoms may have different numbers of neutrons, which are known as isotopes of the element.

The number of protons in the atomic nucleus also determines its electric charge, which in turn determines the electrons of the atom in its non-ionized state. This in turn (by means of the Pauli exclusion principle) determines the atom's various chemical properties. So all carbon atoms, for example, ultimately have identical chemical properties because they all have the same number of protons in their nucleus, and therefore have the same atomic number. It is for this reason that atomic number rather than mass number (or atomic weight) is considered the identifying characteristic of an element.

Atomic Mass

The mass number of an element, A, is the number of nucleons (protons and neutrons) in the atomic nucleus. Different isotopes of a given element are distinguished by their mass numbers, which are conventionally written as a super-index on the left hand side of the atomic symbol (e.g., ^{238}U).

The reltive atomic mass of an element is the average of the atomic masses of all the chemical element's isotopes as found in a particular environment, weighted by isotopic abundance, relative to the *atomic mass unit* (u). This number may be a fraction which is not close to a whole number, due to the averaging process. On the other hand, the atomic mass of a pure isotope is quite close to its mass number. Whereas the mass number is a natural (or whole) number, the atomic mass of a single isotope is a real number which is close to a natural number. In general, it differs slightly from the mass number as the mass of the protons and neutrons is not exactly 1 u, the electrons also contribute slightly to the atomic mass, and because of the nuclear binding energy. For example, the mass of ^{19}F is 18.9984032 u. The only exception to the atomic mass of an isotope not being a natural number is ^{12}C, which has a mass of *exactly* 12, due to the *definition of u* (it is fixed as 1/12th of the mass of a free carbon-12 atom, exactly).

Isotopes

Isotopes are atoms of the same element (that is, with the same number of protons in their atomic nucleus), but having *different* numbers of neutrons. Most (66 of 94) naturally occurring elements have more than one stable isotope. Thus, for example, there are three main isotopes of carbon. All carbon atoms have 6 protons in the nucleus, but they can have either 6, 7, or 8 neutrons. Since the mass numbers of these are 12, 13 and 14 respectively, the three isotopes of carbon are known as carbon-12, carbon-13, and carbon-14, often abbreviated to ^{12}C, ^{13}C, and ^{14}C. Carbon in everyday life and in chemistry is a mixture of ^{12}C, ^{13}C, and ^{14}C atoms.

All three of the isotopes of carbon have the same chemical properties. But they have different nuclear properties. In this example, carbon-12 and carbon-13 are stable atoms, but carbon-14 is unstable; it is slightly radioactive, decaying over time into other elements.

Like carbon, some isotopes of various elements are radioactive and decay into other elements upon radiating an alpha or beta particle. For certain elements, all their isotopes are radioactive isotopes: specifically the elements without any stable isotopes are technetium (atomic number 43), promethium (atomic number 61), and all observed elements with atomic numbers greater than 82.

Of the 80 elements with a stable isotope, 26 have only one stable isotope, and the mean number of stable isotopes for the 80 stable elements is 3.1 stable isotopes per element. The largest number of stable isotopes that occur for an element is 10 (for tin, element 50).

Allotropes

Some elements can be found as multiple elementary substances, known as allotropes, which differ in their structure and properties. For example, carbon can be found as diamond, which has a tetrahedral structure around each carbon atom; graphite, which has layers of carbon atoms with a hexagonal structure stacked on top of each other, graphene, which is a single layer of graphite which is incredibly strong, fullerenes, which have nearly spherical shapes, and carbon nanotubes, which are tubes with a hexagonal structure – there are two types of nanotubes, conducting and semi-conducting. The ability for an element to exist in one of many structural forms is known as 'allotropy'.

Standard State

The standard state, or reference state, of an element is defined as its thermodynamically most stable state at 1 bar at a given temperature (typically at 298.15 K). In thermochemistry, an element is defined to have an enthalpy of formation of zero in its standard state. For example, the reference state for carbon is graphite, because it is more stable than the other allotropes.

Nomenclature

The naming of elements precedes the atomic theory of matter, although at the time it was not known which chemicals were elements and which compounds. When it was learned, existing names (*e.g.*, gold, mercury, iron) were kept in most countries, and national differences emerged over the names of elements either for convenience, linguistic niceties, or nationalism. For example, the Germans use "Wasserstoff" for "hydrogen", "Sauerstoff" for "oxygen" and "Stickstoff" for "nitrogen", while English and some romance languages use "sodium" for "natrium" and "potassium" for "kalium", and the French, Italians, Greeks, Portuguese and Poles prefer "azote/azot/azoto" for "nitrogen".

But for international trade, the official names of the chemical elements both ancient and recent are decided by the International Union of Pure and Applied Chemistry, which has decided on a sort of international English language. That organization has recently prescribed that "aluminium" and "caesium" take the place of the US spellings "aluminum" and "cesium", while the US "sulfur" takes the place of the British "sulphur". But chemicals which are practicable to be sold in bulk within many countries, however, still have national names, and those which do not use the Latin alphabet cannot be expected to use the IUPAC name. According to IUPAC, the full name of an element is not capitalized, even if it is derived from a proper noun such as the elements californium or einsteinium (unless it would be capitalized by some other rule). Isotopes of chemical elements are also uncapitalized if written out: carbon-12 or uranium-235. Symbols of chemical elements, however, are capitalized: thus the symbols for the elements just discussed are Cf and Es; C-12 and U-235.

In the second half of the twentieth century physics laboratories became able to produce nuclei of chemical elements that have a half life too short for them to remain in any appreciable amounts. These are also named by IUPAC, which generally adopts the name chosen by the discoverer. This can lead to the controversial question of which research group actually discovered an element, a question which delayed the naming of elements with atomic number of 104 and higher for a considerable time.

Precursors of such controversies involved the nationalistic namings of elements in the late nineteenth century. For example, *lutetium* was named in reference to Paris, France. The Germans were reluctant to relinquish naming rights to the French, often calling it *cassiopeium*. The British discoverer of *niobium* originally named it *columbium*, in reference to the New World. It was used extensively as such by American publications prior to international standardization.

Specific Chemical Elements

Before chemistry became a science, alchemists had designed arcane symbols for both metals and common compounds. These were however used as abbreviations in diagrams or procedures; there was no concept of atoms combining to form molecules. With his advances in the atomic theory of matter, John Dalton devised his own simpler symbols, based on circles, which were to be used to depict molecules.

The current system of chemical notation was invented by Berzelius. In this typographical system chemical symbols are not used as mere abbreviations - though each consists of letters of the Latin alphabet - they are symbols intended to be used by peoples of all languages and alphabets. The first of these symbols were intended to be fully universal; since Latin was the common language of science at that time, they were abbreviations based on the Latin names of metals - Fe comes from Ferrum, Ag from Argentum. The symbols were not followed by a period (full stop) as abbreviations were. Later chemical elements were also assigned unique chemical symbols, based on the name of the element, but not necessarily in English. For example, sodium has the chemical symbol 'Na' after the Latin *natrium*. The same applies to "W" (wolfram) for tungsten, "Hg" (hydrargyrum) for mercury, "K" (kalium) for potassium, "Au" (aurum) for gold, "Pb" (plumbum) for lead, and "Sb" (stibium) for antimony.

Chemical symbols are understood internationally when element names might need to be translated. There are sometimes differences; for example, the Germans have used "J" instead of "I" for iodine, so the character would not be confused with a roman numeral.

General Chemical Symbols

There are also symbols for series of chemical elements, for comparative formulas. These are one capital letter in length, and the letters are reserved so they are not permitted to be given for the names of specific elements. For example, an "X" is used to indicate a variable group amongst a class of compounds (though usually a halogen), while "R" is used for a radical, meaning a compound structure such as a hydrocarbon chain. The letter "Q" is reserved for "heat" in a chemical reaction. "Y" is also often used as a general chemical symbol, although it is also the symbol of yttrium. "Z" is also frequently used as a general variable group. "L" is used to represent a general ligand in inorganic and organometallic chemistry. "M" is also often used in place of a general metal.

Isotope Symbols

The three main isotopes of the element hydrogen are often written as H for protium, D for deuterium and T for tritium. This is in order to make it easier to use them in chemical equations, as it replaces the need to write out the mass number for each atom. It is written like this:

D_2O (heavy water)

Instead of writing it like this:

^2H_2O

The Periodic Table

The periodic table of the chemical elements is a tabular method of displaying the chemical elements. Although precursors to this table exist, its invention is generally credited to Russian chemist Dmitri Mendeleev in 1869. Mendeleev intended the table to illustrate recurring ("periodic") trends in the properties of the elements. The layout of the table has been refined and extended over time, as new elements have been discovered, and new theoretical models have been developed to explain chemical behavior.

The periodic table is now ubiquitous within the academic discipline of chemistry, providing an extremely useful framework to classify, systematize and compare all the many different forms of chemical behavior. The table has also found wide application in physics, biology, engineering, and industry. The current standard table contains 117 confirmed elements as of March 10, 2009 (while element 118 has been synthesized, element 117 has not).

Abundance

During the early phases of the Big Bang, nucleosynthesis of hydrogen nuclei resulted in the production of hydrogen and helium isotopes, as well as very minuscule amounts (on the order of 10^{-10}) of lithium and beryllium. There is argument about whether or not some boron was produced in the Big Bang, as it has been observed in some very young stars, but no heavier elements than boron were produced. As a result, the primordial abundance of atoms consisted of roughly 75% 1H, 25% 4He, and 0.01% deuterium Subsequent enrichment of galactic halos occurred due to Stellar nucleosynthesis and Supernova nucleosynthesis. However intergalactic space can still closely resemble the primordial abundance, unless it has been enriched by some means.

Table 2.1 shows the twelve most common elements in our galaxy (estimated spectroscopically), as measured in parts per million, by mass. Nearby galaxies that have evolved along similar lines have a corresponding enrichment of elements heavier than hydrogen and helium. The more distant galaxies are being viewed as they appeared in the past, so their abundances of elements appear closer to the primordial mixture. As physical laws and processes appear common throughout the visible universe, however, it is expected that these galaxies will likewise have evolved similar abundances of elements.

Table 2.1: Common elements

Element	Parts per million by mass
Hydrogen	739,000
Helium	240,000
Oxygen	10,400
Carbon	4,600
Neon	1,340
Iron	1,090
Nitrogen	960
Silicon	650
Magnesium	580
Sulfur	440
Potassium	210
Nickel	100

Recently Discovered Element Claims

The first transuranium element (element with atomic number greater than 92) discovered was neptunium in 1940. As of August 2007, only the elements up to 111, roentgenium, have been confirmed as valid by IUPAC, while more or less reliable claims have been made for synthesis of elements 112, 113, 114, 115, 116 and 118. The heaviest element that is believed to have been synthesized to date is element 118, ununoctium, on October 9, 2006, by the Flerov Laboratory of Nuclear Reactions in Dubna, Russia.

Element 117, ununseptium, has not yet been created or discovered, but its place in the periodic table is preestablished.

On April 24, 2008, Amnon Marinov and six other researchers, claimed that element 122 has been detected as a naturally occurring in a thorium deposit. This is the first naturally occurring heavy element in more than 50 years. It has yet to be proved as it is still under confirmation by the university but could be a major development as previously all transuranic elements were artificial.

ISOTOPE RATIO MASS SPECTROMETRY

Isotope ratio mass spectrometry (IRMS) is a specialization of mass spectrometry, in which mass spectrometric methods are used to measure the relative abundance of isotopes in a given sample

Introduction

The isotope ratio mass spectrometer (IRMS) allows the precise measurement of mixtures of stable isotopes This technique has two different applications in the earth and environmental sciences. The analysis of 'stable isotopes' is normally concerned with measuring isotopic variations arising from mass-dependent isotopic fractionation in natural systems. On the other hand, radiogenic isotope analysis involves measuring the abundances of decay-products of natural radioactivity, and is used in most long-lived radiometric dating methods.

Most instruments used for precise determination of isotope ratios are of the magnetic sector type. This type of analyzer is superior to the quadrupole type in this field of research for two reasons. Firstly, it can easily be set up for multiple-collector analysis, and secondly, it gives high-quality 'peak shapes'. Both of these considerations are important for isotope ratio analysis at very high precision and accuracy.

The sector-type instrument designed by Alfred Nier was such an advance in mass spectrometer design that this type of instrument is often called the 'Nier type'. In the most general terms the instrument operates by ionizing the sample of interest, accelerating it over a potential in the kilo-volt range, and separating the resulting stream of ions according to their mass to charge ratio (m/z). The current of each ion beam is then measured using a 'Faraday' detector or multiplier detector.

Many radiogenic isotope measurements are made by ionization of a solid source, whereas stable isotope measurements of light elements (e.g. H, C, O) are usually made in an instrument with a gas source. In the latter case, dual gas inlets enable reliable repetition of measurements by supplying continuous streams of the reference and sample gases, which are sequentially switched by a changeover valve. The IRMS's collector also features an array of Faraday cups (conductive, metal vessels which neutralise ions that hit them whilst themselves becoming charged), or "multicollector", which allows the simultaneous detection of multiple isotopes. Samples must be introduced as pure gases, achieved through combustion, gas chromatographic feeds or chemical trapping. By comparing the detected isotopic ratios to a measured standard, an accurate determination of the isotopic make up of the sample is obtained. For example, carbon isotope ratios are measured relative to the international standard for CO_2. (The CO_2 standard is produced from a fossil belemnite found in the Pee Dee formation, which is a limestone formed in the Cretaceous period in South Carolina, U.S.A.; its ^{13}C:^{12}C ratio is 0.0112372.)

Gas Source ('Stable Isotope') Mass Spectrometry

Instruments have been developed based on several techniques for mass separation and tuned to a wide range of applications. The article below describes one of these application areas, instruments adapted specifically to measure the relative abundance of masses up to around mass number 66.

This field is of interest because the relative variation in mass between isotopes in this range is large enough to give rise to variation in chemical, physical and biological behaviour. This leads to measurable effects on the isotopic composition of samples characteristic of their biological or physical history.

As a specific example, the hydrogen isotope deuterium (heavy hydrogen) is almost double the mass of the common hydrogen isotope. Water molecules containing the common hydrogen isotope (and the common oxygen isotope, mass 16) have a mass of 18. Water incorporating a deuterium atom has a mass of 19, over 5% heavier. The energy to vaporise the heavy water molecule is higher than that to vaporize the normal water so isotope fractionation occurs during the process of evaporation.

Thus a sample of sea water (Vienna Standard Mean Ocean Water, or VSMOW) will exhibit a quite detectable isotopic ratio difference when compared to Arctic snowfall (standard light Arctic precipitation, or SLAP).

It is critical that the sample be processed before entering the mass spectrometer so that only a single chemical species enters at a given time. Generally, samples are combusted or pyrolyzed and the desired species (usually hydrogen gas H_2, nitrogen (N_2), carbon dioxide, or sulfur dioxide) is purified by means of traps, filters, catalysts and/or chromatography.

The two most common types of IRMS instruments are continuous flow and dual inlet. In dual inlet IRMS, purified gas obtained from a sample is alternated rapidly with a standard gas (of known isotopic composition) by means of a system of valves, so that a number of comparison measurements are made of both gases. In continuous flow IRMS, sample preparation occurs immediately before introduction to the IRMS, and the purified gas produced from the sample is measured just once. The standard gas may be measured before and after the sample or after a series of sample measurements. While continuous-flow IRMS instruments can achieve higher sample throughput and are more convenient to use than dual inlet instruments, the yielded data is of approximately 10-fold lower precision.

Static Gas ('Noble Gas') Mass Spectrometry

A static gas mass spectrometer in one in which a gaseous sample for analysis is fed into the source of the instrument and then left in the source without further supply or pumping throughout the analysis. This method can be used for 'stable isotope' analysis of light gases (as above), but it is particularly used in the isotopic analysis of noble gases (= rare or inert gases) for radiometric dating or isotope geochemistry. Important examples are argon-argon dating and helium isotope analysis.

Solid Source ('Radiogenic Isotope') Mass Spectrometry

Several of the isotope systems involved in radiometric dating depend on IRMS using thermal ionization of a solid sample loaded into the source of the mass spectrometer (hence thermal ionization mass spectrometry, TIMS). These methods include rubidium-strontium dating, uranium-lead dating, lead-lead dating and samarium-neodymium dating.

When these isotope ratios are measured by TIMS, mass-dependent fractionation occurs as species are emitted by the hot filament. Fractionation occurs due to the excitation of the sample and therefore must be corrected for accurate measurement of the isotope ratio.

There are several advantages of the TIMS method. It has a simple design, is less expensive than other mass spectrometers, and produces stable ion emissions. It requires a stable power supply, and is suitable for species with a low ionization potential, such as Strontium (Sr), and Lead (Pb).

The disadvantages of this method stem from the maximum temperature achieved in thermal ionization. The hot filament reaches a temperature of less than 2500 degrees Celsius, leading to the inability to create atomic ions of species with a high ionization potential, such as Osmium (Os), and Tungsten (Hf-W). Although the TIMS method can create molecular ions instead in this case, species with high ionization potential can be analyzed more effectively with MC-ICP-MS.

Multiple Collector - Inductively Coupled Plasma - Mass Spectrometry (MC-ICP-MS)

An MC-ICP-MS instrument is a multiple collector mass spectrometer with a plasma source. MC-ICP-MS was developed to improve the precision achievable by ICP-MS during isotope ratio measurements. Conventional ICP-MS analysis uses a quadrupole analyser, which only allows single-collector analysis. Due to the inherent instability of the plasma, this limits the precision of ICP-MS with a quadrupole analyzer to around 1%, which is insufficient for most radiogenic isotope systems.

Isotope ratio analysis for radiometric dating has normally been determined by TIMS. However, some systems (e.g. Hf-W and Lu-Hf) are difficult or impossible to analyse by TIMS, due to the high ionization potential of the elements involved. Therefore these methods can now be analysed using MC-ICP-MS.

Magnetic sector mass spectrometry requires a high acceleration potential in order to properly resolve species with different masses. Since the ion beam is normally positive, a multi-collector ICP-MS (unlike conventional ICP-

MS) needs to have an ion source up to 8000 V positive, in order to keep the analyzer at electrical ground. This was one of the engineering challenges of developing MC-ICP-MS. Magnetic sector mass spectrometry also requires an ion beam with a relatively small range in ion energy. This is achieved in MC-ICP-MS by using an electrostatic and/or quadrupole pre-filter, which allows the ion energy of the beam to be homogenized before entering the magnetic sector.

Moving Wire IRMS (MW-irMS)

Moving wire IRMS is useful for analyzing Carbon-13 ratios of compounds in a solution, such as after purification by liquid chromatography. The solution (or outflow from the chromatography) is dried onto a nickel or stainless steel wire. After the residue is deposited on the wire, it enters a furnace where the sample is converted to CO_2 and water by combustion. The gas stream finally enters a capillary, is dried, ionized, and analyzed This process allows a mixture of compounds to be purified and analyzed continuously, which can decrease the analysis time by a factor of four Moving wire IRMS is quite sensitive, and samples containing as little as 1 nano-mole of Carbon can yield precise (within 1‰) results.

Accelerator Mass Spectrometry

For isotopes occurring at extremely low levels, accelerator mass specrometry (AMS) can be used. For example, the decay rate of th radioisotope ^{14}C is widely used to date organic materials, but his approach is limited to relatively large samples no more than a few thousand years old. AMS extended the range of ^{14}C datig to about 60,000 years BP, and is about $^{10}6$ times more sensitive than conventional IRMS.

Applications

The variation outlined above has applications in hydrology since most samples will lie between these two extremes. Given a sampl of water from an aquifer, and a sufficiently sensitive tool to masure the variation in the isotopic ratio of hydrogen in the saple, it is possible to infer the source, be it ocean water seeping into the aquifer or precipitation seeping into the aquifer, and even to estimate the proportions from each source.

Another application is in paleotemperature measurement for *Paleoclimtology*. For example one technique is based on the variation in isotopic fractionation of oxygen by biological systems with temperature.

Speces of *foraminifera* incorporate oxygen as calcium carbonate in teir shells. The ratio of the oxygen isotopes oxygen 16 and oxygen 18 incorporated into the calcium carbonate varies with temperature and the oxygen isotopic composition of the water. This oxygen remains "fixed" in the calcium carbonate when the forminifera dies, falls to the sea bed, and it's shell becomes part of the sediment. It is possible to select standard species of forminifera from sections through the sediment column, and by mapping the variation in oxygen isotopic ratio, deduce the temperature that the forminifera encountered during life if changes in the oxygen isotopic composition of the water can be constrained.

In forensic science, research suggests that the variation in certain isotope ratios in drugs derived from plant sources (cannabis, cocaine) can be used to determine the drug's continent of origin.

3 Biogeochemistry

The field of biogeochemistry involves scientific study of the chemical, physical, geological, and biological processes and reactions that govern the composition of the natural environment (including the biosphere, the hydrosphere, the pedosphere, the atmosphere, and the lithosphere), and the cycles of matter and energy that transport the Earth's chemical components in time and space. Biogeochemistry is a *systems science* closely related to *systems ecology*.

History

The founder of biogeochemistry is Russian scientist Vladimir Vernadsky, a Russian who, with his 1926 book *The Biosphere*, in the tradition of Mendeleev, is credited with formulating a physics of the earth, as a living whole. Vernadsky distinguished three spheres in the universe domain, where a sphere is a concept similar to the Riemman concept of a space-phase. He observed that each sphere has its own laws of evolution, and that the higher spheres modify and dominate the lowers:

- **Abiotic sphere**: all the non-living energy and material processes.
- **Biosphere**: the life processes that live within the abiotic sphere.
- **Nöesis** or **Nösphere**: the sphere of the cognitive process of man.

Human activities (e.g., agriculture and industry) modify the Biosphere and Abiotic sphere. In the contemporary environment, the amount of influence humans have on the other two spheres is comparable to a geological force.

Early Development of Biogeochemistry

The American limnologist and geochemist G. Evelyn Hutchinson is credited with outlining the broad scope and principles of this new field. More recently, the basic elements of the discipline of biogeochemistry were restated and popularized by the British scientist and writer, James Lovelock, under the label of the *Gaia Hypothesis*. Lovelock emphasizes a concept that life processes regulate the Earth through feedback mechanisms to keep it habitable.

Biogeochemistry Research

There are biogeochemistry research groups in many universities around the world. Since this is a highly inter-disciplinary field, these are situated within a wide range of host disciplines including: atmospheric sciences, biology, ecology, geomicrobiology, environmental chemistry, geology, oceanography and soil science. These are often bracketed into larger disciplines such as earth science and environmental science.

Many researchers investigate the biogeochemical cycles of chemical elements such as carbon, oxygen, nitrogen, phosphorus and sulfur, as well as their stable isotopes. The cycles of trace elements such as the trace metals and the radionuclides are also studied. This research has obvious applications in the exploration for ore deposits and oil, and in remediation of environmental pollution.

Some important research fields for biogeochemistry include:

- modelling of natural systems;
- soil and water acidification recovery processes;
- increased eutrophication of surface waters;
- carbon sequestration;
- soil remediation;
- climate change;
- biogeochemical prospecting for ore deposits.

EUTROPHICATION

Eutrophication is an increase in chemical nutrients — compounds containing nitrogen or phosphorus — in an ecosystem, and may occur on land or in water. However, the term is often used to mean the resultant increase in the ecosystem's primary productivity (excessive plant growth and decay), and further effects including lack of oxygen and severe reductions in water quality, fish, and other animal populations.

Lakes and Rivers

Eutrophication is frequently a result of nutrient pollution, such as the release of sewage effluent, urban stormwater run-off, and run-off carring excess fertilizers into natural waters. However, it may also occur naturally in situations where nutrients accumulate (e.g. depositional environments) or where they flow into systems on an ephemeral basis. Eutrophication generally promotes excessive plant growth and decay, favors certain weedy species over others, and is likely to cause severe reductions in water quality. In aquatic environments, enhanced growth of choking aquatic vegetation or phytoplankton (that is, an algal bloom) disrupts normal functioning of the ecosystem, causing a variety of problems such as a lack of oxygen in the water, needed for fish and shellfish to survive. The water then becomes cloudy, coloured a shade of green, yellow, brown, or red. Human society is impacted as well: eutrophication decreases the resource value of rivers, lakes, and estuaries such that recreation, fishing, hunting, and aesthetic enjoyment are hindered. Health-related problems can occur where eutrophic conditions interfere with drinking water treatment.

Eutrophication was recognized as a pollution problem in European and North American lakes and reservoirs in the mid-20th century Since then, it has become more widespread. Surveys showed that 54% of lakes in Asia are eutrophic; in Europe, 53%; in North America, 48%; in South America, 41%; and in Africa, 28%.

Although eutrophication is commonly caused by human activities, eutrophication can also be a natural process in lakes; thus, eutrophy is a natural condition for many lakes (e.g., in temperate grasslands). Paleolimnologists now recognise that climate change, geology and other external influences are critical in regulating the natural productivity of lakes. Some lakes also demonstrate the reverse process (meiotrophication), becoming less nutrient rich with time.

Eutrophication can also be a natural process in seasonally inundated tropical floodplains such as the Barotse Floodplain of the Zambezi River. The first floodwaters to move down the floodplain after the onset of the rainy season, called "red waters", are usually hypoxic and kill many fish as a result of eutrophication brought on by material picked up by the flood from the plain such as cattle manure, and by the decay of vegetation which grew during the dry season. The process may be made worse by the use of fertilisers in crops such as maize, rice and sugarcane grown on the floodplain.

Human activities can accelerate the rate at which nutrients enter ecosystems. Runoff from agriculture and development, pollution from septic systems and sewers, and other human-related activities increase the flux of both inorganic nutrients and organic substances into terrestrial, and aquatic ecosystems. Elevated atmospheric compounds of nitrogen can increase nitrogen availability.

Phosphorus is often regarded as the main culprit in cases of eutrophication in lakes subjected to point source pollution from sewage. The concentration of algae and the trophic state of lakes correspond well to phosphorus levels in water. Studies conducted in the Experimental Lakes Area in Ontario have shown a relationship between the addition of phosphorus and the rate of eutrophication. Humankind has increased the rate of phosphorus cycling on Earth by four times, mainly due to agricultural fertilizer production and application. Between 1950 and 1995, 600,000,000 tonnes of phosphorus were applied to Earth's surface, primarily on croplands Control of point sources of phosphorus have resulted in rapid control of eutrophication, mainly due to policy changes.

Ocean Waters

Eutrophication is also a common phenomenon in marine, coastal waters. In contrast to freshwater systems, nitrogen is more commonly the key limiting nutrient of marine waters; thus, nitrogen levels have greater importance to understanding eutrophication problems in salt water. Estuaries tend to be naturally eutrophic because land-derived nutrients are concentrated where run-off enters the marine environment in a confined channel. Upwelling in coastal systems also promotes increased productivity by conveying deep, nutrient-rich waters to the surface, where the nutrients can be assimilated by algae.

The World Resources Institute has identified 375 hypoxic coastal zones in the world, concentrated in coastal areas in Western Europe, the Eastern and Sout.

In addition to runoff from land, atmospheric anthropogenic fixed nitrogen can enter the open ocean. A study in 2008 found that this could account for around one third of the ocean's external (non-recycled) nitrogen supply and up to three per cent of the annual new marine biological production It has been suggested that accumulating reactive nitrogen in the environment may prove as serious as putting carbon dioxide in the atmosphere.

Terrestrial Ecosystems

Although traditionally thought of as enrichment of aquatic systems by addition of fertilizers into lakes, bays, or other semi-enclosed waters (even slow-moving rivers), terrestrial ecosystems are subject to similarly adverse impacts Increased content of nitrates in soil frequently leads to undesirable changes in vegetation composition and many plant species are endangered as a result of eutrophication in terrestrial ecosystems, e.g. the majority of orchid species in Europe. Ecosystems (like some meadows, forests and bogs that are characterized by low nutrient content and species-rich, slowly growing vegetation adapted to lower nutrient levels) are overgrown by faster growing and more competitive species-poor vegetation, like tall grasses, that can take advantage of unnaturally elevated nitrogen level and the area may be changed beyond recognition and vulnerable species may be lost. Eg. species-rich fens are overtaken by reed or reedgrass species, spectacular forest undergrowth affected by run-off from nearby fertilized field is turned into a thick nettle and bramble shrub.

Chemical forms of nitrogen are most often of concern with regard to eutrophication because plants have high nitrogen requirements so that additions of nitrogen compounds stimulate plant growth (primary production). This is also the case with increased levels of phosphorus. Nitrogen is not readily available in soil because N_2, a gaseous form of nitrogen, is very stable and unavailable directly to higher plants. Terrestrial ecosystems rely on microbial nitrogen fixation to convert N_2 into other physical forms (such as nitrates). However, there is a limit to how much nitrogen can be utilized. Ecosystems receiving more nitrogen than the plants require are called nitrogen-saturated. Saturated terrestrial ecosystems contribute both inorganic and organic nitrogen to freshwater, coastal, and marine eutrophication, where nitrogen is also typically a limiting nutrient. However, because phosphorus is generally much less soluble than nitrogen, it is leached from the soil at a much slower rate than nitrogen. Consequently, phosphorus is much more important as a limiting nutrient in aquatic systems.

Ecological Effects

Many ecological effects can arise from stimulating primary production, but there are three particularly troubling ecological impacts: decreased biodiversity, changes in species composition and dominance, and toxicity effects.

- Increased biomass of phytoplankton.
- Toxic or inedible phytoplankton species
- Increases in blooms of gelatinous zooplankton
- Decreased biomass of benthic and epiphytic algae
- Changes in macrophyte species composition and biomass
- Decreases in water transparency (increased turbidity)
- Colour, smell, and water treatment problems
- Dissolved oxygen depletion
- Increased incidences of fish kills
- Loss of desirable fish species
- Reductions in harvestable fish and shellfish
- Decreases in perceived aesthetic value of the water body

Decreased Biodiversity

When an ecosystem experiences an increase in nutrients, primary producers reap the benefits first. In aquatic ecosystems, species such as algae experience a population increase (called an algal bloom). Algal blooms limit the sunlight available to bottom-dwelling organisms and cause wide swings in the amount of dissolved oxygen in the water. Oxygen is required by all respiring plants and animals and it is replenished in daylight by photosynthesizing plants and algae. Under eutrophic conditions, dissolved oxygen greatly increases during the day, but is greatly reduced after dark by the respiring algae and by microorganisms that feed on the increasing mass of dead algae. When dissolved oxygen levels decline to hypoxic levels, fish and other marine animals suffocate. As a result, creatures such as fish, shrimp, and especially immobile bottom dwellers die off. In extreme cases, anaerobic conditions ensue, promoting growth of bacteria such as *Clostridium botulinum* that produces toxins deadly to birds and mammals. Zones where this occurs are known as dead zones.

New Species Invasion

Eutrophication may cause competitive release by making abundant a normally limiting nutrient. This process causes shifts in the species composition of ecosystems. For instance, an increase in nitrogen might allow new, competitive species to invade and out-compete original inhabitant species. This has been shown to occur in New England salt marshes.

Toxicity

Some algal blooms, otherwise called "nuisance algae" or "harmful algal blooms", are toxic to plants and animals. Toxic compounds they produce can make their way up the food chain, resulting in animal mortality. Freshwater algal blooms can pose a threat to livestock. When the algae die or are eaten, neuro- and hepatotoxins are released which can kill animals and may pose a threat to humans. An example of algal toxins working their way into humans is the case of shellfish poisoning. Biotoxins created during algal blooms are taken up by shellfish (mussels, oysters), leading to these human foods acquiring the toxicity and poisoning humans. Examples include paralytic, neurotoxic, and diarrhoetic shellfish poisoning. Other marine animals can be vectors for such toxins, as in the case of ciguatera, where it is typically a predator fish that accumulates the toxin and then poisons humans. Nitrogen can also cause toxic effects directly. When this nutrient is leached into groundwater, drinking water can be affected because concentrations of nitrogen are not filtered out. 'Blue baby syndrome, or methaemoglobinaemia is caused when nitrate (NO_3) contaminated water is given to human babies. The anaerobic environment in their stomach causes the nitrates to be converted into nitrites (NO_2). Nitrite binds to hemoglobin to form methemoglobin, a form that does not carry oxygen. The baby essentially suffocates as its body doesn't receive enough oxygen. The problem does not happen in breast fed babies.

Sources of High Nutrient Runoff

In order to gauge how to best prevent eutrophication from occurring, specific sources that contribute to nutrient loading must be identified. There are two common sources of nutrients and organic matter: point and nonpoint sources.

Point Sources

Point sources are directly attributable to one influence. In point sources the nutrient waste travels directly from source to water. Point sources are relatively easy to regulate.

Non-point Sources

Non-point source pollution (also known as 'diffuse' or 'runoff' pollution) is that which comes from ill-defined and diffuse sources. Nonpoint sources are difficult to regulate and usually vary spatially and temporally (with season, precipitation, and other irregular events).

It has been shown that nitrogen transport is correlated with various indices of human activity in watersheds, including the amount of development Agriculture and development are activities that contribute most to nutrient loading. There are three reasons that nonpoint sources are especially troublesome.

Soil Retention

Nutrients from human activities tend to accumulate in soils and remain there for years. It has been shown that the amount of phosphorus lost to surface waters increases linearly with the amount of phosphorus in the soil. Thus much of the nutrient loading in soil eventually makes its way to water. Nitrogen, similarly, has a turnover time of decades or more.

Runoff to Surface Water and Leaching to Groundwater

Nutrients from human activities tend to travel from land to either surface or ground water. Nitrogen in particular is removed through storm drains, sewage pipes, and other forms of surface runoff. Nutrient losses in runoff and leachate are often associated with agriculture. Modern agriculture often involves the application of nutrients onto fields in order to maximise production. However, farmers frequently apply more nutrients than are taken up by crops or pastures. Regulations aimed at minimising nutrient exports from agriculture are typically far less stringent than those placed on sewage treatment plants and other point source polluters.

Atmospheric Deposition

Nitrogen is released into the air because of ammonia volatilization and nitrous oxide production. The combustion of fossil fuels is a large human-initiated contributor to atmospheric nitrogen pollution. Atmospheric deposition (e.g., in the form of acid rain) can also affect nutrient concentration in water especially in highly industrialized regions.

Other Causes

Any factor that causes increased nutrient concentrations can potentially lead to eutrophication. In modeling eutrophication, the rate of water renewal plays a critical role; stagnant water is allowed to collect more nutrients than bodies with replenished water supplies. It has also been shown that the drying of wetlands causes an increase in nutrient concentration and subsequent eutrophication blooms.

Prevention and Reversal

Eutrophication poses a problem not only to ecosystems, but to humans as well. Reducing eutrophication should be a key concern when considering future policy, and a sustainable solution for everyone, including farmers and ranchers, seems feasible. While eutrophication does pose problems, humans should be aware that natural runoff (which causes algal blooms in the wild) is common in ecosystems and should thus not reverse nutrient concentrations beyond normal levels.

Ceanup measures have been mostly, but not completely, successful. Finnish phosphorus removal measures started in the mid-1970s and have targeted rivers and lakes polluted by industrial and municipal discharges. These efforts have had a 90% removal efficiency. Still, some targeted point sources did not show a decrease in runoff despite reduction efforts.

Minimizing Nonpoint Pollution: Future Work

Nonpoint pollution is the most difficult source of nutrients to manage. The literature suggests, though, that when these sources are controlled, eutrophication decreases. The following steps are recommended to minimize the amount of pollution that can enter aquatic ecosystems from ambiguous sources.

Riparian Buffer Zones

Studies show that intercepting non-point pollution between the source and the water is a successful means of prevention. Riparian buffer zones are interfaces between a flowing body of water and land, and have been created near waterways in an attempt to filter pollutants; sediment and nutrients are deposited here instead of in water. Creating buffer zones near farms and roads is another possible way to prevent nutrients from traveling too far. Still, studies have shown that the effects of atmospheric nitrogen pollution can reach far past the buffer zone. This suggests that the most effective means of prevention is from the primary source.

Prevention Policy

Laws regulating the discharge and treatment of sewage have led to dramatic nutrient reductions to surrounding ecosystems, but it is generally agreed that a policy regulating agricultural use of fertilizer and animal waste must be imposed. In Japan the amount of nitrogen produced by livestock is adequate to serve the fertilizer needs for the agriculture industry Thus, it is not unreasonable to command livestock owners to clean up animal waste — which when left stagnant will leach into ground water.

Nitrogen Testing and Modeling

Soil Nitrogen Testing (N-Testing) is a technique that helps farmers optimize the amount of fertilizer applied to crops. By testing fields with this method, farmers saw a decrease in fertilizer application costs, a decrease in nitrogen lost to surrounding sources, or both By testing the soil and modeling the bare minimum amount of fertilizer needed, farmers reap economic benefits while the environment remains clean and pure.

Organic Farming

Researchers at the National Academy of Sciences found that organically fertilizing fields "significantly reduce harmful nitrate leaching" over conventionally fertilized fields.

NATURAL STATE OF ALGAL BLOOMS

Although the intensity, frequency and extent of algal blooms has tended to increase in response to human activity and human-induced eutrophication, algal blooms are a naturally-occurring phenomenon. The rise and fall of algae populations, as with the population of other living things, is a feature of a healthy ecosystem. Rectification actions aimed at abating eutrophication and algal blooms are usually desirable, but the focus of intervention should not necessarily be aimed at eliminating blooms, but towards creating a sustainable balance that maintains or improves ecosystem health.

SYSTEMS ECOLOGY

Systems ecology is an interdisciplinary field of ecology, taking a holistic approach to the study of ecological systems, especially ecosystems. Systems ecology can be seen as an application of general systems theory to ecology. Central to the systems ecology approach is the idea that an ecosystem is a complex system exhibiting emergent properties. Systems ecology focuses on interactions and transactions within and between biological and ecological systems, and is especially concerned with the way the functioning of ecosystems can be influenced by human interventions. It uses and extends concepts from thermodynamics and develops other macroscopic descriptions of complex systems.

Overview

Systems ecology seeks a holistic view of the interactions and transactions within and between biological and ecological systems. Systems ecologists realise that the function of any ecosystem can be influenced by human economics in fundamental ways. They have therefore taken an additional transdisciplinary step by including economics in the consideration of ecological-economic systems. In the words of R.L. Kitching:

- Systems ecology can be defined as the approach to the study of ecology of organisms using the techniques and philosophy of systems analysis: that is, the methods and tools developed, largely in engineering, for studying, characteriszing and making predictions about complex entities, that is, systems.

- In any study of an ecological system, an essential early procedure is to draw a diagram of the system of interest . . . diagrams indicate the system's boundaries by a solid line. Within these boundaries, series of components are isolated which have been chosen to represent that portion of the world in which the systems analyst is interested . . . If there are no connections across the systems' boundaries with the surrounding systems environments, the systems are described as closed. Ecological work, however, deals almost exclusively with open systems.

As a mode of scientific enquiry, a central feature of Systems Ecology is the general application of the principles of energetics to all systems at any scale. Perhaps the most notable proponent of this view was Howard T. Odum - sometimes considered the father of ecosystems ecology. In this approach the principles of energetics constitute ecosystem principles. Reasoning by formal analogy from one system to another enables the Systems Ecologist to see principles functioning in an analogous manner across system-scale boundaries. H.T. Odum commonly used the Energy Systems Language as a tool for making systems diagrams and flow charts.

The fourth of these principles, the principle of maximum power efficiency, takes central place in the analysis and synthesis of ecological systems. The fourth principle suggests that the most evolutionarily advantageous system function occurs when the environmental load matches the internal resistance of the system. The further the environmental load is from matching the internal resistance, the further the system is away from its sustainable steady state. Therefore the systems ecologist engages in a task of resistance and impedance matching in ecological engineering, just as the electronic engineer would do.

Closely Related Fields

Deep Ecology

Deep ecology is a school of philosophy pioneered by the Norwegian Philosopher, Gandhian scholar and environmental activist Arne Naess. Created in 1973 at an environmental conference in Budapest, it argues that the school of environmental management is anthropocentric, that the natural environment is not only "more complex than we imagine, it is more complex than we can imagine" Concerned with the development of an "ecological self", which views the human ego as a part of a living system, rather than apart from such systems, "Experiential Deep Ecology" of Joanna Macy, John Seed and others, seeks to transcend altruism with a deeper self-interest, based upon biospherical equality beyond human chauvinism.

Earth Systems Engineering and Management

Earth systems engineering and management (ESEM) is a discipline used to analyze, design, engineer and manage complex environmental systems. It entails a wide range of subject areas including anthroplogy, engineering, environmental science, ethics and philosophy. At its core, ESEM looks to "rationally design and manage coupled human-natural systems in a highly integrated and ethical fashion."

Ecological Economics

Ecological economics is a transdisciplinary field of academic research that addresses the dynamic and spatial interdependence between human economies and natural ecosystems. Ecological economics brings together and connects different disciplines, within the natural and social sciences but especially between these broad areas. As the name suggests, the field is made up of researchers with a background in economics and ecology. An important motivation for the emergence of ecological economics has been criticism on the assumptions and approaches of traditional (mainstream) environmental and resource economics.

Ecological Energetics

Ecological energetics is the quantitative study of the flow of energy through ecological systems. It aims to uncover the principles which describe the propensity of such energy flows through the trophic, or 'energy availing' levels of ecological networks. In systems ecology the principles of ecosystem energy flows or "ecosystem laws" (i.e. principles of ecological energetics) are considered formally analogous to the principles of energetics.

Ecological Humanities

Ecological humanities aims to bridge the divides between the sciences and the humanities, and between Western, Eastern and Indigenous ways of knowing nature. Like ecocentric political theory, the ecological humanities are characterised by a connectivity ontology and a commitment to two fundamental axioms relating to the need to submit to ecological laws and to see humanity as part of a larger living system.

Ecosystem Ecology

Ecosystem ecology is the integrated study of biotic and abiotic components of ecosystems and their interactions within an ecosystem framework. This science examines how ecosystems work and relates this to their components such as chemicals, bedrock, soil, plants, and animals. Ecosystem ecology examines physical and biological structure and examines how these ecosystem characteristics interact.

The relationship between systems ecology and ecosystem ecology is complex. Much of systems ecology can be considered a subset of ecosystem ecology. Ecosystem ecology also utilizes methods that have little to do with the holistic approach of systems ecology. However, systems ecology more actively considers external influences such as economics that usually fall outside the bounds of ecosystem ecology. Whereas ecosystem ecology can be defined as the scientific study of ecosystems, systems ecology is more of a particular approach to the study of ecological systems and phenomena that interact with these systems.

Industrial Ecology

Industrial ecology is the study of industrial process as linear (open loop) systems, in which resource and capital investments move through the system to become waste, to a closed loop system where wastes become inputs for new processes.

4 Cosmochemistry

Cosmochemistry or chemical cosmology is concerned with the origin and development of the elements and their isotopes, primarily within the Solar System. Cosmochemistry is that branch of chemistry with astronomy which deals with the study of origin and development of elements and their isotopes,primarily within the solar system. The term was coined by Harold Urey Closely related fields are astrochemistry, a branch of astronomy concerned with measuring chemical elements in other parts of our Galaxy and in other galaxies; astrophysics, which includes the study of physical processes in stars, supernovas, and our own solar system that may result in chemical and isotopic changes; and planetary science, of which cosmochemistry is in large part a subdiscipline.

Cosmochemistry frequently involves direct measurement of physical samples in laboratories on the Earth. Samples are most often meteorites and micrometeorites, which include material that originated on the Moon, Mars, many different asteroids, and quite possibly comets, as well as with samples returned from the Moon by manned and robotic missions. Other materials available for direct study by cosmochemists are tiny particles from Comet Wild 2 returned to Earth in 2006 by NASA's Stardust mission, and samples of solar wind, captured by the Genesis mission and returned to Earth in 2004. The most exotic available samples are tiny presolar grains that are embedded in some primitive chondritic meteorites, which originated elsewhere in the galaxy prior to the formation of the solar system. Similar interstellar dust particles may also have been collected by the Stardust mission, although this has not yet been confirmed.

NASA has funded a research programme in Cosmochemistry since the 1970s

COSMOLOGY

Cosmology (from Greek *kosmos*, "universe"; and - *-logia*, "study") is study of the Universe in its totality, and by extension, humanity's place in it. Though the word *cosmology* is recent (first used in 1730 in Christian Wolff's *Cosmologia Generalis*), study of the Universe has a long history involving science, philosophy, esotericism, and religion.

Disciplines

The physics and astrophysics have to play a central role in shaping the understanding of the universe through scientific observation and experiment; or what is known as physical cosmology shaped through both mathematics and observation in the analysis of the whole universe. In other words, in this discipline, which focuses on the universe as it exists on the largest scale and at the earliest moments, is generally understood to begin with the big bang (possibly combined with cosmic inflation) - an expansion of space from which the Universe itself is thought to have emerged ~13.7 $\pm 0.2 \times 10^9$ (13.7 billion) years ago From its violent beginnings and until its various speculative ends, cosmologists propose that the history of the Universe has been governed entirely by physical laws. Theories of an impersonal universe governed by physical laws were first proposed by Roger Bacon, a somewhat persecuted member of the Catholic Church. Between the domains of religion and science, stands the philosophical perspective of metaphysical cosmology. This ancient field of study seeks to draw intuitive conclusions about the nature of the universe, man, god and/or their relationships based on the extension of some set of presumed facts borrowed from spiritual experience and/or observation.

But metaphysical cosmology has also been observed as the placing of man in the universe in relationship to all other entities. This is demonstrated by the observation made by Marcus Aurelius of a man's place in that relationship: " He who does not know what the world is does not know where he is, and he who does not know for what purpose the world exists, does not know who he is, nor what the world is. This is the purpose of the ancient metaphysical cosmology. However, Stoicism rejected Aristotle's theory of universals as being "in the things themselves," calling them "figments of the mind." Stanford Encyclopedia of Philosophy adopting the concept of universals as being "concepts," and therefore of the mind, and therefore controllable by free will. Thus, we get the analysis of Aurelius' that the nature of the universe is not from "intuition," but from a free-will, conceptual understanding of the nature of the universe.

Cosmology is often an important aspect of the creation myths of religions that seek to explain the existence and nature of reality. In some cases, views about the creation (cosmogony) and destruction (eschatology) of the universe play a central role in shaping a framework of religious cosmology for understanding humanity's role in the universe.

A more contemporary distinction between religion and philosophy, esoteric cosmology is distinguished from religion in its less tradition-bound construction and reliance on modern "intellectual understanding" rather than faith, and from philosophy in its emphasis on spirituality as a formative concept.

There are many historical cosmologies:

"...the universe itself acts on us as a random, inefficient, and yet in the long run effective, teaching machine. ... our way of looking at the universe has gradually evolved through a natural selection of ideas." —Steven Weinberg

Historical Cosmologies

Historical Descriptions of the Cosmos

Name	Author and date	Classification	Remarks
1	2	3	4
Babylonian cosmology	Babylonian literature (1900-1200 BC)	Plurality of heavens and earth	The earth and heavens are a "spatial whole, even one of round shape," revolving around the "cult-place of the deity" rather than the Earth, and three is a plurality of heavens and earths.

(Contd...)

1	2	3	4
Brahmanda (Hindu Cosmology)	Hindu Rigveda (1500-1200 B.C.)	Cyclical or oscillating, Infinite in time	The universe sustains for around 311 trillion and 40 billion years that is 100 years of the cosmic creator Brahma. There is a smaller period of unmanifestation in around 4 billion years that is one day in the life of Brahma. The universe cycles between expansion and total collapse. After one cycle of the life of Brahma another universe follows up to an infinite number each of which exists for a time period of 311 trillion and 40 billion years. It also speaks of an infinite number of universes at one given point of time. Universe expanded from a concentrated form—a point called a Bindu. The universe, as a living entity, is bound to the perpetual cycle of birth, death and rebirth.

(Contd…)

1	2	3	4
Atomist universe	Anaxagoras (500-428 B.C.) and later Epicurus	Infinite in extent	The universe contains only two things: an infinite number of tiny seeds, or atoms, and the void of infinite extent. All atoms are made of the same substance, but differ in size and shape. Objects are formed from atom aggregations and decay back into atoms. Incorporates Leucippus' principle of causality: "nothing happens at random; everything happens out of reason and necessity." The universe was not ruled by gods.
Pythagorean Universe	Philolaus (d. 390 BC)	Existence of a "Central Fire" at the centre of the Universe	At the centre of the Universe is a central fire, around which the Earth, Sun, Moon and Planets revolve uniformly. The sun revolves around the central fire once a year, the stars are immobile. The earth in its motion maintains the same hidden face towards the central fire, hence it is never seen. This is the first known non-geocentric model of the Universe.

(Contd...)

1	2	3	4
Stoic universe	Stoics (300 B.C.-200 A.D.)	Island universe	The cosmos is finite and surrounded by an infinite void. It is in a state of flux, as it pulsates in size and periodically passes through upheavals and conflagrations
Aristotelian universe	Aristotle (384-322 B.C.)	Geocentric, static, steady state, finite extent, infinite time	Spherical earth is surrounded by concentric celestial spheres. Universe exists unchanged throughout eternity. Contains a 5th element called aether (later known as quintessence).
Aristarchean universe	Aristarchus (circa 280 B.C.)	Heliocentric	Earth rotates daily on its axis and revolves annually about the sun in a circular orbit. Sphere of fixed stars in centred about the sun.
Seleucian universe	Seleucus of Seleucia (circa 190 B.C.)	Heliocentric	Modifications to the Aristarchean universe, with the inculsion of the ride phenomenon to explain heliocentrism.

(Contd…)

1	2	3	4
Ptolemaic model (based on Aristotelian universe	Ptolemy (2nd century A.D.)	Geocentric	Universe orbits about a stationary Earth. Planets move in circular epicycles, each having a centre that moved in a larger circular orbit (called an eccentric or a deferent) around a centre-point near the Earth. The use of equants added another level of complexity and allowed astronomers to predict the positions of the planets. The most successful universe model of all time, using the criterion of longevity. Almagest (the Great System).
Aryabhatan model	Aryabhata (499)	Geocentric or Heliocentric	The Earth rotates and the planets move in elliptical orbits, possibly around either the earth or the sun. It is uncertain whether the model is geocentric or heliocentric due to planetary, orbits given with respect to both the earth and the sun.

(Contd…)

1	2	3	4
Abrahamic universe	Medieval philosophers (500-1200)	Finite in time	The universe that is finite in time and has a beginning is proposed by the Christian philosopher, John Philoponus, who argues against the ancient Greek notion of an infinite past. Logical arguments supporting a finite universe are developed by the early Muslim philosopher, Alkindus; the Jewish philosopher, Saadia Gaon; and the Muslim theologian, Algazel.
Multiversal cosmology	Fakhr al-Din al-Razi (1149-1209)	Multiverse, multiple worlds and universes	Al-Razi rejects the idea of an Earth-centred universe and, in the context of his commentary on the Qur'anic verse "All praise belongs to God, Lord of the Worlds," proposes that the universe has more than "a thousand thousand worlds beyond this world such that each one of those worlds be bigger and more massive than this world as well as having the like of what this world has," and that there could be an infinite number of universes.

(Contd…)

1	2	3	4
Maragha models	Maragha school (1259-1528)	Geocentric	Various modifications to Ptolemaic model and Aristotelian universe, including rejection of equant and eccentrics at Maragheh observatory, and introduction of Tusi couple by Al-Tusi. Alternative models later proposed, including the first accurate lunar model by Ibn al-Shatir, a model rejecting stationary Earth in favour of Earth's rotation by Ali Kuscu, and planetary model incorporating "circular inertia" by Al-Birjandi.
Nilakanthan model	Nilakantha Somayaji (1444-1544)	Geocentric and Heliocentric	A universe in which the planets orbit the Sun and the Sun orbits the Earth, similar to the later Tychonic system.
Copernican universe	Nicolaus Copernicus (1543)	Heliocentric	The geocentric Maragha model of Ibn al-Shatir adapted to meet the requirements of the ancient heliocentric Aristarchean universe in his *De revolutionibus orbium coelestium.*
Tychonic system	Tycho Brahe (1546-1601)	Geocentric and Heliocentric	A universe in which the planets orbit the Sun and the Sun orbits the Earth, similar to the earlier Nilakanthan model.

(Contd…)

1	2	3	4
Static Newtonian	Sir Isaac Newton (1642-1727)	Static (evolving), steady state, infinite	Every particle in the universe attracts every other particle. Matter on the large scale is uniformly distributed Gravitationally balanced but unstable.
Cartesian Vortex universe	René Descartes 17th century	Static (evolving) steady, state, infinite	A system of huge swirling whirlpools of aethereal or fine matter produces what we would call gravitational effects. His vacuum was not empty. All space was filled with matter that swirled around in large and small vortices.
Hierarchical universe	Immanuel Kant, Johann Lambert 1700s	Static (evolving), steady state, infinite	Matter is clustered on ever larger scales of hierarchy. Matter is endlessly being recycled.
Einstein universe with a cosmological constant	Albert Einstein 1917	Static (nominally) Bounded (finite)	"Matter without motion." Contains uniformly distributed matter. Uniformly curved spherical space; based on Riemann's hypersphere. Curvature is set equal to ^. In effect ^ is equivalent to a repulsive force which counteracts gravity. Unstable.

(Contd…)

1	2	3	4
De Sitter universe	William de Sitter 1917	Expanding flat space Steady state. ^ > 0	"Motion without matter." Only apparently static. Based on Einstein's General Relativity. Space expands with constant acceleration. Scale factor (radius of universe) increases exponentially, i.e. constant inflation.
MacMillan	William MacMillan 1920s	Static and steady state	New matter is created from radiation Starlight is perpetually recycled into new matter particles.
Friedmann universe of spherical space	Alexander Friedmann 1922	Spherical expanding space k = +1; no ^	Positive curvature. Curvature constant k = +1 Expands then recollapses. Spatially closed (finite)
Friedmann universe of hyperbolic space	Alexander Friedmann 1924	Hyperbolic expanding space k = -1; no ^	Negative curvature. Said to be infinite (but ambiguous). Unbounded. Expands forever
Dirac large number hypothesis	Paul Dirac 1930s	Expanding	Demands a large variation in G, which decreases with time. Gravity weakens as universe evolves.

(Contd…)

1	2	3	4
Friedmann zero-curvature aka the Einstein-DeSitter universe	Einstein & DeSitter 1932	Expanding flat space $k = 0$; ^ = 0 Critical density	Curvature constant $k = 0$. Said to be infinite (but ambiguous). 'Unbounded cosmos of limited extent.' Expands forever. 'Simplest' of all known universes. Named after but not considered by Friedmann. Has a deceleration term $q = ½$ which means that its expansion rate slows down.
Georges Lemaître the original Big Bang.aka Friedmann-Lemaître Model	Georges Lemaître 1927-29	Expansion ^ > 0 ^ > |Gravity|	^ is positive and has a magnitude greater than Gravity. Universe has initial high density state ('primeval atom'). Followed by a two stage expansion. ^ is used to destabilize the universe (Lemaître is considered to be the father of the big bang model.)
Oscillating universe (aka Friedmann-Einstein; was latter's 1st choice after rejecting his own 1917 model	Favoured by Friedmann 1920s	Expanding and contracting in cycles	Time is endless and beginningless; thus avoids the beginning-of-time paradox. Perpetual cycles of big bang followed by big crunch.

(Contd...)

1	2	3	4
Eddington	Arthur Eddington 1930	First Static then expands	Static Einstein 1917 universe with its instability disturbed into expansion mode; with relentless matter dilution becomes a DeSitter universe. ^ dominates gravity.
Milne universe of kinematic relativity	Edward Milne 1933, 1935; William H. McCrea, 1930s	Kinematic expansion with No space expansion	Rejects general relativity and the expanding space paradigm. Gravity not included as initial assumption. Obeys cosmological principle and rules of special relativity. The Milne expanding universe consists of a finite spherical cloud of particles (or galaxies) that expands within flat space which is infinite and otherwise empty. It has a centre and a cosmic edge (the surface of the particle cloud) which expands at light speed. His explanation of gravity was elaborated and unconvincing. For instance, his universe has an infinite number of particles, hence infinite mass, within a finite cosmic volume.

(Contd…)

1	2	3	4
Friedmann-Lemaître-Walker class of models	Howard Robertson, Arthur Walker, 1935	Uniformly expanding	Class of universes that are homogenous and isotropic. Spacetime separates into uniformly curved space and cosmic time common to all comoving observers. The formulation system is now known as the FLRW or Robertson-Walker metrics of cosmic time and curved space.
Steady-state expanding (Bondi & Gold)	Herman Bondi Thomas Gold 1948	Expanding, steady state, infinite	Matter creation rate maintains constant density. Continuous creation out of nothing from nowhere. Exponential expansion. Deceleration term $q = -1$.
Steady-state expanding (Hoyle)	Fred Hoyle 1948	Expanding, steady state; but unstable	Matter creation rate maintains constant density. But since matter creation rate must be exactly balanced with the space expansion rate the system is unstable.
Ambiplasma	Hannes Alfvén 1965 Oskar Klein	Cellular universe, expanding by means of matter-animatter annihilation	Based on the concept of plasma cosmology. The universe is viewed as *meta-galaxies* divided by double layers —hence its bubble-like nature. Other universes are formed from other bubbles. Ongoing cosmic matter-animatter annihilations keep the bubbles separated and moving apart preventing them from interacting.

(Contd…)

1	2	3	4
Brans-Dicke	Carl H. Brans; Robert H. Dicke	Expanding	Based on Mach's principles. G. Varies with time as universe expands. "But nobody is quite sure what Mach's principle actually means.
Cosmic inflation	Alan Guth 1980	Big Bang with modification to solve horizon problem and flatness problem	Based on the concept of hot inflation. The universe is viewed as a multiple quantum flux—hence its bubble-like nature. Other universes are formed from other bubbles. Ongoing cosmic expansion kept the bubbles separated and moving apart preventing them from interacting.
Eternal Inflation (a multiple universe model)	Andreï Linde 1983	Big Bang with cosmic inflation	A multiverse, based on the concept of cold inflation, in which inflationary event occur at random each with independent initial conditions; some expand into bubble universes supposedly like our entire cosmos. Bubbles nucleate in a spacetime foam.

(Contd...)

1	2	3	4
Cyclic model	Paul Steinhardt; Neil Turok 2002	Expanding and contracting in cycles; M theory	Two parallel orbifold planes or M-branes collide periodically in a higher dimensional space. With quintessence of dark energy.
Cyclic model	Lauris Baum; Paul Frampton 2007	Solution of Tolman's entropy problem	Phantom dark energy fragments universe into large number of disconnected patches. Our patch contracts containing only dark energy with zero entropy.

Note: The term "static" simply means not expanding and not contracting. Symbol G represents Newton's gravitational constant; * (Lambda) is the cosmological constant.

Physical Cosmology

Physical cosmology is the branch of physics and astrophysics that deals with the study of the physical origins and evolution of the Universe. It also includes the study of the nature of the Universe on its very largest scales. In its earliest form it was what is now known as celestial mechanics, the study of the heavens. The Greek philosophers Aristarchus of Samos, Aristotle and Ptolemy proposed different cosmological theories. In particular, the geocentric Ptolemaic system was the accepted theory to explain the motion of the heavens until Nicolaus Copernicus, and subsequently Johannes Kepler and Galileo Galilei proposed a heliocentric system in the 16th century. This is known as one of the most famous examples of epistemological rupture in physical cosmology.

With Isaac Newton and the 1687 publication of *Principia Mathematica*, the problem of the motion of the heavens was finally solved. Newton provided a physical mechanism for Kepler's laws and his law of universal gravitation allowed the anomalies in previous systems, caused by gravitational interaction between the planets, to be resolved. A fundamental difference between Newton's cosmology and those preceding it was the Copernican principle that the bodies on earth obey the same physical laws as all the celestial bodies. This was a crucial philosophical advance in physical cosmology.

Modern scientific cosmology is usually considered to have begun in 1917 with Albert Einstein's publication of his final modification of general relativity in the paper "Cosmological Considerations of the General Theory of Relativity," (although this paper was not widely available outside of Germany until the end of World War I). General relativity prompted cosmogonists such as Willem de Sitter, Karl Schwarzschild and Arthur Eddington to explore the astronomical consequences of the theory, which enhanced the growing ability of astronomers to study very distant objects. Prior to this (and for some time afterwards), physicists assumed that the Universe was static and unchanging. In parallel to this dynamic approach to cosmology, a debate was unfolding regarding the nature of the cosmos itself. On the one hand, Mount Wilson astronomer Harlow Shapley championed the model of a cosmos made up of the Milky Way star system only. Heber D. Curtis, on the other hand, suggested spiral nebulae were star systems in their own right, island universes. This difference of ideas came to a climax with the organization of the Great Debate at the meeting of the (US) National Academy of Sciences in Washington on 26 April 1920. The resolution of the debate on the structure of the cosmos came with the detection of novae in the Andromeda galaxy by Edwin Hubble in 1923 and 1924. Their distance established spiral nebulae well beyond the edge of the Milky Way and as galaxies of their own. Subsequent modeling of the universe explored the possibility that the cosmological constant introduced by Einstein in his 1917 paper may result in an expanding universe, depending on its value. Thus the big bang theory was proposed by the Belgian priest Georges Lemaître in 1927 which was subsequently corroborated by Edwin Hubble's discovery of the red shift in 1929 and later by the discovery of the cosmic microwave background radiation by Arno Penzias and Robert Woodrow Wilson in 1964. These findings were a first step to rule out some of many alternative physical cosmologies.

Observations made by the COBE and WMAP satellites observing this background radiation have effectively, in many scientists' eyes, transformed cosmology from a highly speculative science into a predictive science, as these observations matched predictions made by a theory called Cosmic inflation, which is a modification of the standard big bang theory. This has led many to refer to modern times as the "Golden age of cosmology".

Metaphysical Cosmology

In philosophy and metaphysics, cosmology deals with the world as the totality of space, time and all phenomena. Historically, it has had quite a broad scope, and in many cases was founded in religion. The ancient Greeks did not draw a distinction between this use and their model for the cosmos. However, in modern use it addresses questions about the Universe which are beyond the scope of science. It is distinguished from religious cosmology in that it approaches these questions using philosophical methods (e.g. dialectics). Modern metaphysical cosmology tries to address questions such as:

- What is the origin of the Universe? What is its first cause? Is its existence necessary? (see monism, pantheism, emanationism and creationism).
- What are the ultimate material components of the Universe? (see mechanism, dynamism, hylomorphism, atomism).
- What is the ultimate reason for the existence of the Universe? Does the cosmos have a purpose? (see teleology).

Religious Cosmology

Many world religions have creation myths that explain the beginnings of the Universe and life. Often these are derived from scriptural teachings and held to be part of the faith's dogma, but in some cases these are also extended through the use of philosophical and metaphysical arguments.

In some creation myths, the universe was created by a direct act of a god or gods who are also responsible for the creation of humanity (see creationism). In many cases, religious cosmologies also foretell the end of the Universe, either through another divine act or as part of the original design.

- Both Christianity and Judaism rely on the Genesis narrative as a scriptural account of cosmology.
- Islam relies on understanding from the Qur'an as its major source for explaining cosmology.
- Certain adherents of Buddhism, Hinduism and Jainism believe that the Universe passes through endless cycles of creation and destruction, each cycle lasting for trillions of years (e.g. 331 trillion years, or the life-span of Brahma, according to Hinduism), and each cycle with sub-cycles of local creation and destruction (e.g. 4.32 billion years, or a day of Brahma, according to Hinduism). The Vedic (Hindu) view of the world sees one true divine principle self-projecting as the divine word, 'birthing' the cosmos that we know from the monistic *Hiranyagarbha* or Golden Womb.
- A complex mixture of native Vedic gods, spirits, and demons, overlaid with imported Hindu and Buddhist deities, beliefs, and practices are the key to the Sri Lankan cosmology.
- The Australian Aboriginal concept of *Dreaming* explains the creation of the universe as an eternal continuum; *everywhen*. Through certain ceremonies, the past "opens up" and comes into the present. Each topographical feature is a manifestation of dormant creation spirits; each individual has personal Dreamings and ceremonial responsibilities to look after the spirits/land, determined at birth, within this belief framework.

Many religions accept the findings of physical cosmology, in particular the Big Bang, and some, such as the Roman Catholic Church, have embraced it as suggesting a philosophical first cause. Others have tried to use the methodology of science to advocate for their own religious cosmology, as in intelligent design or creationist cosmologies.

Esoteric Cosmology

Many esoteric and occult teachings involve highly elaborate cosmologies. These constitute a "map" of the Universe and of states of existences and consciousness according to the worldview of that particular doctrine. Such cosmologies cover many of the same concerns also addressed by religious and philosophical cosmology, such as the origin, purpose, and destiny of the Universe and of consciousness and the nature of existence. For this reason it is difficult to distinguish where religion or philosophy end and esotericism and/or occultism begins.

Common themes addressed in esoteric cosmology are emanation, involution, evolution, epigenesis, planes of existence, hierarchies of spiritual beings, cosmic cycles (e.g., cosmic year, Yuga), yogic or spiritual disciplines, and references to altered states of consciousness. Examples of esoteric cosmologies can be found in modern Theosophy, Gnosticism, Tantra (especially Kashmir Shaivism), *Kabbalah*, or *Sufism*.

5 Earth

Earth is the third planet from the Sun. Earth is the largest of the terrestrial planets in the Solar System in diameter, mass and density. It is also referred to as *the World* and *Terra*.

Home to millions of species, including humans, Earth is the only place in the universe where life is known to exist. The planet formed 4.54 billion years ago, and life appeared on its surface within a billion years. Since then, Earth's biosphere has significantly altered the atmosphere and other abiotic conditions on the planet, enabling the proliferation of aerobic organisms as well as the formation of the ozone layer which, together with Earth's magnetic field, blocks harmful radiation, permitting life on land. The physical properties of the Earth, as well as its geological history and orbit, allowed life to persist during this period. The world is expected to continue supporting life for another 1.5 billion years, after which the rising luminosity of the Sun will eliminate the biosphere.

Earth's outer surface is divided into several rigid segments, or tectonic plates, that gradually migrate across the surface over periods of many millions of years. About 71% of the surface is covered with salt-water oceans, the remainder consisting of continents and islands; liquid water, necessary for all known life, is not known to exist on any other planet's surface. Earth's interior remains active, with a thick layer of relatively solid mantle, a liquid outer core that generates a magnetic field, and a solid iron inner core.

Earth interacts with other objects in outer space, including the Sun and the Moon. At present, Earth orbits the Sun once for every roughly 366.26 times it rotates about its axis. This length of time is a sidereal year, which is equal to 365.26 solar days. The Earth's axis of rotation is tilted 23.4° away from the perpendicular to its orbital plane, producing seasonal variations on the planet's surface with a period of one tropical year (365.24 solar days). Earth's only known natural satellite, the Moon, which began orbiting it about 4.53 billion years ago, provides ocean tides, stabilizes the axial tilt and gradually slows the planet's rotation. Between approximately 4.1 and 3.8 billion years ago, asteroid impacts during the Late Heavy Bombardment caused significant changes to the surface environment.

Both the mineral resources of the planet, as well as the products of the biosphere, contribute resources that are used to support a global human population. The inhabitants are grouped into about 200 independent sovereign states, which interact through diplomacy, travel, trade and military action. Human cultures have developed many views of the planet, including personification as a deity, a belief in a flat Earth, and a modern perspective of the world as an integrated environment that requires stewardship.

Chronology

Scientists have been able to reconstruct detailed information about the planet's past. The earliest dated solar system material is dated to 4.5672 ± 0.0006 billion years ago, and by 4.54 billion years ago (within an uncertainty of 1%) the Earth and the other planets in the Solar System formed out of the solar nebula—a disk-shaped mass of dust and gas left over from the formation of the Sun. This assembly of the Earth through accretion was largely completed within 10-20 million years Initially molten, the outer layer of the planet Earth cooled to form a solid crust when water began accumulating in the atmosphere. The Moon formed soon afterward, most likely as the result of a Mars-sized object (sometimes called Theia) with about 10% of the Earth's mass impacting the Earth in a glancing blow. Some of this object's mass would have merged with the Earth and a portion would have been ejected into space, but enough material would have been sent into orbit to form the Moon.

Outgassing and *volcanic* activity produced the primordial atmosphere. Condensing water vapor, augmented by ice and liquid water delivered by asteroids and the larger proto-planets, comets, and trans-Neptunian objects produced the oceans. Two major models have been proposed for the rate of continental growth: steady growth to the present-day and rapid growth early in Earth history. Current research shows that the second option is most likely, with rapid initial growth of continental crust followed by a long-term steady continental area On time scales lasting hundreds of millions of years, the surface continually reshaped itself as continents formed and broke up. The continents migrated across the surface, occasionally combining to form a supercontinent. Roughly 750 million years ago (mya), one of the earliest known supercontinents, Rodinia, began to break apart. The continents later recombined to form Pannotia, 600-540 mya, then finally Pangaea, which broke apart 180 mya.

Evolution of Life

At present, Earth provides the only example of an environment that can sustain the evolution of life Highly energetic chemistry is believed to have produced a self-replicating molecule around 4 billion years ago, and half a billion years later the last common ancestor of all life existed. The development of photosynthesis allowed the Sun's energy to be harvested directly by life forms; the resultant oxygen accumulated in the atmosphere and formed in a layer of ozone (a form of molecular oxygen [O_3]) in the upper atmosphere. The incorporation of smaller cells within larger ones resulted in the development of complex cells called eukaryotes True multicellular organisms formed as cells within colonies became increasingly specialized. Aided by the absorption of harmful ultraviolet radiation by the ozone layer, life colonized the surface of Earth.

Since the 1960s, it has been hypothesized that severe glacial action between 750 and 580 mya, during the Neoproterozoic, covered much of the planet in a sheet of ice. This hypothesis has been termed "Snowball Earth", and is of particular interest because it preceded the Cambrian explosion, when multicellular life forms began to proliferate.

Following the Cambrian explosion, about 535 mya, there have been five mass extinctions. The last extinction event was 65 mya, when a meteorite collision probably triggered the extinction of the (non-avian) dinosaurs and other large reptiles, but spared small animals such as mammals, which then resembled shrews. Over the past 65 million years, mammalian life has diversified, and several mya, an African ape-like animal gained the ability to stand upright This enabled tool use and encouraged communication that provided the nutrition and stimulation needed for a larger brain. The development of agriculture, and then civilization, allowed humans to influence the Earth in a short time span as no other life form had, affecting both the nature and quantity of other life forms.

The present pattern of *ice ages* began about 40 mya and then intensified during the Pleistocene about 3 mya. The polar regions have since undergone repeated cycles of glaciation and thaw, repeating every 40-100,000 years. The last ice age ended 10,000 years ago.

Future

The future of the planet is closely tied to that of the Sun. As a result of the steady accumulation of helium ash at the Sun's core, the star's total luminosity will slowly increase. The luminosity of the Sun will grow by 10 percent over the next 1.1 Gyr (1.1 billion years) and by 40% over the next 3.5 Gyr Climate models indicate that the rise in radiation reaching the Earth is likely to have dire consequences, including the possible loss of the planet's oceans

The Earth's increasing surface temperature will accelerate the inorganic CO_2 cycle, reducing its concentration to the lethal levels for plants (10 ppm for C_4 photosynthesis) in 900 million years. The lack of vegetation will result in the loss of oxygen in the atmosphere, so animal life will become extinct within several million more years But even if the Sun were eternal and stable, the continued internal cooling of the Earth would have resulted in a loss of much of its atmosphere and oceans due to reduced volcanism After another billion years all surface water will have disappeared and the mean global temperature will reach 70°C The Earth is expected to be effectively habitable for about another 500 million years.

The Sun, as part of its evolution, will become a red giant in about 5 Gyr. Models predict that the Sun will expand out to about 250 times its present size, roughly 1 AU (150,000,000 km) Earth's fate is less clear. As a red giant, the Sun will lose roughly 30% of its mass, so, without tidal effects, the Earth will move to an orbit 1.7 AU (250,000,000 km) from the Sun when the star reaches it maximum radius. Therefore, the planet is expected to escape envelopment by the expanded Sun's sparse outer atmosphere, though most, if not all, remaining life will be destroyed because of the Sun's increased luminosity However, a more recent simulation indicates that Earth's orbit will decay due to tidal effects and drag, causing it to enter the red giant Sun's atmosphere and be destroyed.

Composition and Structure

Earth is a terrestrial planet, meaning that it is a rocky body, rather than a gas giant like Jupiter. It is the largest of the four solar terrestrial planets, both in terms of size and mass. Of these four planets, Earth also has the highest density, the highest surface gravity, the strongest magnetic field, and fastest rotation It also is the only terrestrial planet with active plate tectonics.

Shape

The shape of the Earth is very close to that of an oblate spheroid, a sphere squished along the orientation from pole to pole such that there is a bulge around the equator This bulge results from the rotation of the Earth, and causes the diameter at the equator to be 43 km larger than the pole to pole diameter The average diameter of the reference spheroid is about 12,742 km, which is approximately 40,000 km/p, as the meter was originally defined as 1/10,000,000 of the distance from the equator to the North Pole through Paris, France.

Local topography deviates from this idealized spheroid, though on a global scale, these deviations are very small: Earth has a tolerance of about one part in about 584, or 0.17%, from the reference spheroid, which is less than the 0.22% tolerance allowed in billiard balls. The largest local deviations in the rocky surface of the Earth are Mount Everest (8,848 m above local sea level) and the Mariana Trench (10,911 m below local sea level). Because of the equatorial bulge, the feature farthest from the center of the Earth is actually Mount Chimborazo in Ecuador.

Chemical Composition

The mass of the Earth is approximately 5.98×10^{24} kg. It is composed mostly of iron (32.1%), oxygen (30.1%), silicon (15.1%), magnesium (13.9%), sulfur (2.9%), nickel (1.8%), calcium (1.5%), and aluminium (1.4%); with the remaining 1.2% consisting of trace amounts of other elements. Due to mass segregation, the core region is believed to be primarily composed of iron (88.8%), with smaller amounts of nickel (5.8%), sulfur (4.5%), and less than 1% trace elements.

Table 5.1: Chemical Composition of the Crust

Compound	Formula	Composition
silica	SiO_2	59.71%
alumina	Al_2O_3	15.41%
lime	CaO	4.90%
Magnesia	MgO	4.36%
sodium oxide	Na_2O	3.55%
iron(II) oxide	FeO	3.52%
potassium oxide	K_2O	2.80%
iron(III) oxide	Fe_2O_3	2.63%
water	H_2O	1.52%
titanium dioxide	TiO_2	0.60%
phosphorus pentoxide	P_2O_5	0.22%
Total		99.22%

The geochemist F. W. Clarke calculated that a little more than 47% of the Earth's crust consists of oxygen. The more common rock constituents of the Earth's crust are nearly all oxides; chlorine, sulfur and fluorine are the only important exceptions to this and their total amount in any rock is usually much less than 1%. The principal oxides are silica, alumina, iron oxides, lime, magnesia, potash and soda. The silica functions principally as an acid, forming silicates, and all the commonest minerals of igneous rocks are of this nature. From a computation based on 1,672 analyses of all kinds of rocks, Clarke deduced that 99.22% were composed of 11 oxides (see the table at right.) All the other constituents occur only in very small quantities.

Internal Structure

The interior of the Earth, like that of the other terrestrial planets, is divided into layers by their chemical or physical (rheological) properties. The outer layer of the Earth is a chemically-distinct silicate solid crust, which is underlain by a highly viscous solid mantle. The crust is separated from the mantle by the Mohorovicic discontinuity, and the thickness of the crust varies: averaging 6 km under the oceans and 30-50 km on the continents. The crust and the cold, rigid, top of the upper mantle are collectively known as the lithosphere, and it is of the lithosphere that the tectonic plates are comprised. Beneath the lithosphere is the asthenosphere, a relatively low-viscosity layer on which the lithosphere rides. Important changes in crystal structure within the mantle occur at 410 and 660 kilometers below the surface, spanning a transition zone that separates the upper and lower mantle. Beneath the mantle, an extremely low viscosity liquid outer core lies above a solid inner core. The inner core may rotate at a slightly higher angular velocity than the remainder of the planet, advancing by 0.1–0.5° per year.

Heat

The internal heat of the planet is from a combination residual heat from planetary accretion (20%) and heat produced through radioactive decay (80%) The major heat-producing isotopes in the Earth are potassium-40, uranium-238, uranium-235, and thorium-232. At the center of the planet, the temperature may be up to 7,000 K and the pressure could reach 360 GPa. Because much of the heat is provided by radioactive decay, scientists believe that early in Earth history, before isotopes with short half-lives had been depleted, Earth's heat production would have been much higher. This extra heat production, twice present-day at approximately three billion years ago, would have increased temperature gradients within the Earth, increasing the rates of mantle convection and plate tectonics, and allowing the production of igneous rocks such as komatiites that are not formed today.

Total heat loss from the earth is 4.2×10^{13} Watts. A portion of the core's thermal energy is transported toward the crust by Mantle plumes; a form of convection consisting of upwellings of higher-temperature rock. These plumes can produce hotspots and flood basalts. More of the heat in the Earth is lost through plate tectonics, by mantle upwelling associated with mid-ocean ridges. The final major mode of heat loss is through conduction through the lithosphere, majority of which occurs in the oceans due to the crust there being much thinner than that of the continents.

Tectonic Plates

The mechanically rigid outer layer of the Earth, the lithosphere, is broken into pieces called tectonic plates. These plates are rigid segments that move in relation to one another at one of three types of plate boundaries: Convergent boundaries, at which two plates come together, Divergent boundaries, at which two plates are pulled apart, and Transform boundaries, in which two plates slide past one another laterally. Earthquakes, volcanic activity, mountain-building, and oceanic trench formation can occur along these plate boundaries. The tectonic plates ride on top of the asthenosphere, the solid but less-viscous part of the upper mantle that can flow and move along with the plates, and their motion is strongly coupled with patterns convection inside the Earth's mantle.

As the tectonic plates migrate across the planet, the ocean floor is subducted under the leading edges of the plates at convergent boundaries. At the same time, the upwelling of mantle material at divergent boundaries creates mid-ocean ridges. The combination of these processes continually recycles the oceanic crust back into the mantle. Because of this recycling, most of the ocean floor is less than 100 million years in age. The oldest oceanic crust is located in the Western Pacific, and has an estimated age of about 200 million years. By comparison, the oldest dated continental crust is 4030 million years old.

Other notable plates include the Indian Plate, the Arabian Plate, the Caribbean Plate, the Nazca Plate off the west coast of South America and the Scotia Plate in the southern Atlantic Ocean. The Australian Plate actually fused with Indian Plate between 50 and 55 million years ago. The fastest-moving plates are the oceanic plates, with the Cocos Plate advancing at a rate of 75 mm/yr and the Pacific Plate moving 52-69 mm/yr. At the other extreme, the slowest-moving plate is the Eurasian Plate, progressing at a typical rate of about 21 mm/yr.

Surface

The Earth's terrain varies greatly from place to place. About 70.8% of the surface is covered by water, with much of the continental shelf below sea level. The submerged surface has mountainous features, including a globe-spanning mid-ocean ridge system, as well as undersea volcanoes, oceanic trenches, submarine canyons, oceanic plateaus and abyssal plains. The remaining 29.2% not covered by water consists of mountains, deserts, plains, plateaus, and other geomorphologies.

The planetary surface undergoes reshaping over geological time periods due to the effects of tectonics and erosion. The surface features built up or deformed through plate tectonics are subject to steady weathering from precipitation, thermal cycles, and chemical effects. Glaciation, coastal erosion, the build-up of coral reefs, and large meteorite impacts also act to reshape the landscape.

The continental crust consists of lower density material such as the igneous rocks granite and andesite. Less common is basalt, a denser volcanic rock that is the primary constituent of the ocean floors Sedimentary rock is formed from the accumulation of sediment that becomes compacted together. Nearly 75% of the continental surfaces are covered by sedimentary rocks, although they form only about 5% of the crust. The third form of rock material found on Earth is metamorphic rock, which is created from the transformation of pre-existing rock types through high pressures, high temperatures, or both. The most abundant silicate minerals on the Earth's surface include quartz, the feldspars, amphibole, mica, pyroxene and olivine. Common carbonate minerals include calcite (found in limestone), aragonite and dolomite.

The pedosphere is the outermost layer of the Earth that is composed of soil and subject to soil formation processes. It exists at the interface of the lithosphere, atmosphere, hydrosphere and biosphere. Currently the total arable land is 13.31% of the land surface, with only 4.71% supporting permanent crops. Close to 40% of the Earth's land surface is presently used for cropland and pasture, or an estimated 1.3 × 107 km^2 of cropland and 3.4 × 10^7 km^2 of pastureland.

The elevation of the land surface of the Earth varies from the low point of -418 m at the Dead Sea, to a 2005-estimated maximum altitude of 8,848 m at the top of Mount Everest. The mean height of land above sea level is 840 m.

The abundance of water on Earth's surface is a unique feature that distinguishes the "Blue Planet" from others in the Solar System. The Earth's hydrosphere consists chiefly of the oceans, but technically includes all water surfaces in the world, including inland seas, lakes, rivers, and underground waters down to a depth of 2,000 m. The deepest underwater location is Challenger Deep of the Mariana Trench in the Pacific Ocean with a depth of -10,911.4 m The average depth of the oceans is 3,800 m, more than four times the average height of the continents.

The mass of the oceans is approximately 1.35 × 10^{18} metric tons, or about 1/4400 of the total mass of the Earth, and occupies a volume of 1.386×10^9 km^3. If all of the land on Earth were spread evenly, water would rise to an altitude of more than 2.7 km About 97.5% of the water is saline, while the remaining 2.5% is fresh water. The majority of the fresh water, about 68.7%, is currently in the form of ice.

About 3.5‰ of the total mass of the oceans consists of salt. Most of this salt was released from volcanic activity or extracted from cool, igneous rocks. The oceans are also a reservoir of dissolved atmospheric gases, which are essential for the survival of many aquatic life forms. Sea water has an important influence on the world's climate, with the oceans acting as a large heat reservoir. Shifts in the oceanic temperature distribution can cause significant weather shifts, such as the El Niño-Southern Oscillation.

Atmosphere

The atmospheric pressure on the surface of the Earth averages 101.325 kPa, with a scale height of about 8.5 km. It is 78% nitrogen and 21% oxygen, with trace amounts of water vapour, carbon dioxide and other gaseous molecules. The height of the troposphere varies with latitude, ranging between 8 km at the poles to 17 km at the equator, with some variation due to weather and seasonal factors.

Earth's biosphere has significantly altered its atmosphere. Oxygenic photosynthesis evolved 2.7 billion years ago, forming the primarily nitrogen-oxygen atmosphere that exists today. This change enabled the proliferation of aerobic organisms as well as the formation of the ozone layer which, together with Earth's magnetic field, blocks ultraviolet solar radiation, permitting life on land. Other atmospheric functions important to life on Earth include transporting water vapor, providing useful gases, causing small meteors to burn up before they strike the surface, and moderating temperature This last phenomenon is known as the greenhouse effect: trace molecules within the atmosphere serve to capture thermal energy emitted from the ground, thereby raising the average temperature. Carbon dioxide, water vapor, methane and ozone are the primary greenhouse gases in the Earth's atmosphere. Without this heat-retention effect, the average surface temperature would be -18 °C and life would likely not exist.

Weather and Climate

The Earth's atmosphere has no definite boundary, slowly becoming thinner and fading into outer space. Three-quarters of the atmosphere's mass is contained within the first 11 km of the planet's surface. This lowest layer is called the troposphere. Energy from the Sun heats this layer, and the surface below, causing expansion of the air. This lower density air then rises, and is replaced by cooler, higher density air. The result is atmospheric circulation that drives the weather and climate through redistribution of heat energy.

The primary atmospheric circulation bands consist of the trade winds in the equatorial region below 30° latitude and the westerlies in the mid-latitudes between 30° and 60° Ocean currents are also important factors in determining climate, particularly the thermohaline circulation that distributes heat energy from the equatorial oceans to the polar regions.

Water vapour generated through surface evaporation is transported by circulatory patterns in the atmosphere. When atmospheric conditions permit an uplift of warm, humid air, this water condenses and settles to the surface as precipitation. Most of the water is then transported back to lower elevations by river systems, usually returning to the oceans or being deposited into lakes. This water cycle is a vital mechanism for supporting life on land, and is a primary factor in the erosion of surface features over geological periods. Precipitation patterns vary widely, ranging from several meters of water per year to less than a millimeter. Atmospheric circulation, topological features and temperature differences determine the average precipitation that falls in each region.

The Earth can be sub-divided into specific latitudinal belts of approximately homogeneous climate. Ranging from the equator to the polar regions, these are the tropical (or equatorial), subtropical, temperate and polar climates.

Climate can also be classified based on the temperature and precipitation, with the climate regions characterized by fairly uniform air masses. The commonly used Köppen climate classification system (as modified by Wladimir Köppen's student Rudolph Geiger) has five broad groups (humid tropics, arid, humid middle latitudes, continental and cold polar), which are further divided into more specific subtypes.

Upper Atmosphere

Above the troposphere, the atmosphere is usually divided into the stratosphere, mesosphere, and thermosphere. Each of these layers has a different lapse rate, defining the rate of change in temperature with height. Beyond these, the exosphere thins out into the magnetosphere. This is where the Earth's magnetic fields interact with the solar wind. An important part of the atmosphere for life on Earth is the ozone layer, a component of the stratosphere that partially shields the surface from ultraviolet light. The Kármán line, defined as 100 km above the Earth's surface, is a working definition for the boundary between atmosphere and space.

Due to thermal energy, some of the molecules at the outer edge of the Earth's atmosphere have their velocity increased to the point where they can escape from the planet's gravity. This results in a slow but steady leakage of the atmosphere into space. Because unfixed hydrogen has a low molecular weight, it can achieve escape velocity more readily and it leaks into outer space at a greater rate than other gasses The leakage of hydrogen into space is a contributing factor in pushing the Earth from an initially reducing state to its current oxidizing one. Photosynthesis provided a source of free oxygen, but the loss of reducing agents such as hydrogen is believed to have been a necessary precondition for the widespread accumulation of oxygen in the atmosphere Hence the ability of hydrogen to escape from the Earth's atmosphere may have influenced the nature of life that developed on the planet. In the current, oxygen-rich atmosphere most hydrogen is converted into water before it has an opportunity to escape. Instead, most of the hydrogen loss comes from the destruction of methane in the upper atmosphere.

Magnetic Field

The Earth's magnetic field is shaped roughly as a magnetic dipole, with the poles currently located proximate to the planet's geographic poles. According to dynamo theory, the field is generated within the molten outer core region where heat creates convection motions of conducting materials, generating electric currents. These in turn produce the Earth's magnetic field. The convection movements in the core are chaotic in nature, and periodically change alignment. This results in field reversals at irregular intervals averaging a few times every million years. The most recent reversal occurred approximately 700,000 years ago.

The field forms the magnetosphere, which deflects particles in the solar wind. The sunward edge of the bow shock is located at about 13 times the radius of the Earth. The collision between the magnetic field and the solar wind forms the Van Allen radiation belts, a pair of concentric, torus-shaped regions of energetic charged particles. When the plasma enters the Earth's atmosphere at the magnetic poles, it forms the aurora.

Orbit and Rotation

Rotation

Earth's rotation period relative to the Sun—its mean solar day—is 86,400 seconds of mean solar time. Each of these seconds is slightly longer than an SI second because Earth's solar day is now slightly longer than it was during the 19th century due to tidal acceleration.

Earth's rotation period relative to the fixed stars, called its *stellar day* by the International Earth Rotation and

Reference Systems Service (IERS), is 86164.098903691 seconds of mean solar time (UT1), or $23^h\ 56^m\ 4.098903691^s$. Earth's rotation period relative to the precessing or moving mean vernal equinox, misnamed its *sidereal day*, is 86164.09053083288 seconds of mean solar time (UT1) ($23^h\ 56^m\ 4.09053083288^s$). Thus the sidereal day is shorter than the stellar day by about 8.4 ms. The length of the mean solar day in SI seconds is available from the IERS for the periods 1623-2005 and 1962-2005.

Apart from meteors within the atmosphere and low-orbiting satellites, the main apparent motion of celestial bodies in the Earth's sky is to the west at a rate of 15°/h = 15'/min. This is equivalent to an apparent diameter of the Sun or Moon every two minutes; the apparent sizes of the Sun and the Moon are approximately the same.

Orbit

Earth orbits the Sun at an average distance of about 150 million kilometers every 365.2564 mean solar days, or one sidereal year. From Earth, this gives an apparent movement of the Sun eastward with respect to the stars at a rate of about 1°/day, or a Sun or Moon diameter every 12 hours. Because of this motion, on average it takes 24 hours—a solar day—for Earth to complete a full rotation about its axis so that the Sun returns to the meridian. The orbital speed of the Earth averages about 30 km/s (108,000 km/h), which is fast enough to cover the planet's diameter (about 12,600 km) in seven minutes, and the distance to the Moon (384,000 km) in four hours.

The Moon revolves with the Earth around a common barycenter every 27.32 days relative to the background stars. When combined with the Earth–Moon system's common revolution around the Sun, the period of the synodic month, from new moon to new moon, is 29.53 days. Viewed from the celestial north pole, the motion of Earth, the Moon and their axial rotations are all counter-clockwise. Viewed from a vantage point above the north poles of both the Sun and the Earth, the Earth appears to revolve in a counterclockwise direction about the Sun. The orbital and axial planes are not precisely aligned: Earth's axis is tilted some 23.5 degrees from the perpendicular to the Earth–Sun plane, and the Earth–Moon plane is tilted about 5 degrees against the Earth-Sun plane. Without this tilt, there would be an eclipse every two weeks, alternating between lunar eclipses and solar eclipses.

The Hill sphere, or gravitational sphere of influence, of the Earth is about 1.5 Gm (or 1,500,000 kilometers) in radius This is maximum distance at which the Earth's gravitational influence is stronger than the more distant Sun and planets. Objects must orbit the Earth within this radius, or they can become unbound by the gravitational perturbation of the Sun.

Earth, along with the Solar System, is situated in the Milky Way galaxy, orbiting about 28,000 light years from the center of the galaxy, and about 20 light years above the galaxy's equatorial plane in the Orion spiral arm.

Axial Tilt and Seasons

Because of the axial tilt of the Earth, the amount of sunlight reaching any given point on the surface varies over the course of the year. This results in seasonal change in climate, with summer in the northern hemisphere occurring when the North Pole is pointing toward the Sun, and winter taking place when the pole is pointed away. During the summer, the day lasts longer and the Sun climbs higher in the sky. In winter, the climate becomes generally cooler and the days shorter. Above the Arctic Circle, an extreme case is reached where there is no daylight at all for part of the year—a polar night. In the southern hemisphere the situation is exactly reversed, with the South Pole oriented opposite the direction of the North Pole.

By astronomical convention, the four seasons are determined by the solstices—the point in the orbit of maximum axial tilt toward or away from the Sun—and the equinoxes, when the direction of the tilt and the direction to the Sun are perpendicular. Winter solstice occurs on about December 21, summer solstice is near June 21, spring equinox is around March 20 and autumnal equinox is about September 23.

The angle of the Earth's tilt is relatively stable over long periods of time. However, the tilt does undergo nutation; a slight, irregular motion with a main period of 18.6 years. The orientation (rather than the angle) of the Earth's axis also changes over time, precessing around in a complete circle over each 25,800 year cycle; this precession is the reason for the difference between a sidereal year and a tropical year. Both of these motions are caused by the varying attraction of the Sun and Moon on the Earth's equatorial bulge. From the perspective of the Earth, the poles also migrate a few meters across the surface. This polar motion has multiple, cyclical components, which collectively are termed quasiperiodic motion. In addition to an annual component to this motion, there is a 14-month cycle called the Chandler wobble. The rotational velocity of the Earth also varies in a phenomenon known as length of day variation.

In modern times, Earth's perihelion occurs around January 3, and the aphelion around July 4. However, these dates change over time due to precession and other orbital factors, which follow cyclical patterns known as Milankovitch cycles. The changing Earth-Sun distance results in an increase of about 6.9% in solar energy reaching the Earth at perihelion relative to aphelion. Since the southern hemisphere is tilted toward the Sun at about the same time that the Earth reaches the closest approach to the Sun, the southern hemisphere receives slightly more energy from the Sun than does the northern over the course of a year. However, this effect is much less significant than the total energy change due to the axial tilt, and most of the excess energy is absorbed by the higher proportion of water in the southern hemisphere.

Moon

The Moon is a relatively large, terrestrial, planet-like satellite, with a diameter about one-quarter of the Earth's. It is the largest moon in the Solar System relative to the size of its planet. (Charon is larger relative to the dwarf planet Pluto.) The natural satellites orbiting other planets are called "moons" after Earth's Moon.

The gravitational attraction between the Earth and Moon causes tides on Earth. The same effect on the Moon has led to its tidal locking: its rotation period is the same as the time it takes to orbit the Earth. As a result, it always presents the same face to the planet. As the Moon orbits Earth, different parts of its face are illuminated by the Sun, leading to the lunar phases; the dark part of the face is separated from the light part by the solar terminator.

Because of their tidal interaction, the Moon recedes from Earth at the rate of approximately 38 mm a year. Over millions of years, these tiny modifications—and the lengthening of Earth's day by about 23 µs a year—add up to significant changes During the Devonian period, for example, (approximately 410 million years ago) there were 400 days in a year, with each day lasting 21.8 hours.

The Moon may have dramatically affected the development of life by moderating the planet's climate. Paleontological evidence and computer simulations show that Earth's axial tilt is stabilized by tidal interactions with the Moon Some theorists believe that without this stabilization against the torques applied by the Sun and planets to the Earth's equatorial bulge, the rotational axis might be chaotically unstable, exhibiting chaotic changes over millions of years, as appears to be the case for Mars. If Earth's axis of rotation were to approach the plane of the ecliptic, extremely severe weather could result from the resulting extreme seasonal differences. One pole would be pointed directly toward the Sun during *summer* and directly away during *winter*. Planetary scientists who have studied the effect claim that this might kill all large animal and higher plant life However, this is a controversial subject, and further studies of Mars—which has a similar rotation period and axial tilt as Earth, but not its large Moon or liquid core—may settle the matter.

Viewed from Earth, the Moon is just far enough away to have very nearly the same apparent-sized disk as the Sun. The angular size (or solid angle) of these two bodies match because, although the Sun's diameter is about 400 times as large as the Moon's, it is also 400 times more distant. This allows total and annular eclipses to occur on Earth.

The most widely accepted theory of the Moon's origin, the giant impact theory, states that it formed from the collision of a Mars-size protoplanet called Theia with the early Earth. This hypothesis explains (among other things) the Moon's relative lack of iron and volatile elements, and the fact that its composition is nearly identical to that of the Earth's crust.

Earth has at least two co-orbital asteroids, 3753 Cruithne and 2002 AA_{29}.

Habitability

A planet that can sustain life is termed habitable, even if life did not originate there. The Earth provides the (currently understood) requisite conditions of liquid water, an environment where complex organic molecules can assemble, and sufficient energy to sustain metabolism. The distance of the Earth from the Sun, as well as its orbital eccentricity, rate of rotation, axial tilt, geological history, sustaining atmosphere and protective magnetic field all contribute to the conditions necessary to originate and sustain life on this planet.

Biosphere

The planet's life forms are sometimes said to form a "biosphere". This biosphere is generally believed to have begun evolving about 3.5 billion years ago. Earth is the only place in the universe where life is known to exist. Some scientists believe that Earth-like biospheres might be rare.

The biosphere is divided into a number of biomes, inhabited by broadly similar plants and animals. On land primarily latitude and height above the sea level separates biomes. Terrestrial biomes lying within the Arctic, Antarctic Circle or in high altitudes are relatively barren of plant and animal life, while the greatest latitudinal diversity of species is found at the Equator.

Natural Resources and Land Use

The Earth provides resources that are exploitable by humans for useful purposes. Some of these are non-renewable resources, such as mineral fuels, that are difficult to replenish on a short time scale.

Large deposits of fossil fuels are obtained from the Earth's crust, consisting of coal, petroleum, natural gas and methane clathrate. These deposits are used by humans both for energy production and as feedstock for chemical production. Mineral ore bodies have also been formed in Earth's crust through a process of Ore genesis, resulting from actions of erosion and plate tectonics. These bodies form concentrated sources for many metals and other useful elements.

The Earth's biosphere produces many useful biological products for humans, including (but far from limited to) food, wood, pharmaceuticals, oxygen, and the recycling of many organic wastes. The land-based ecosystem depends upon topsoil and fresh water, and the oceanic ecosystem depends upon dissolved nutrients washed down from the land Humans also live on the land by using building materials to construct shelters. In 1993, human use of land is approximately.

6 Rock (Geology)

In geology, rock is a naturally occurring solid aggregate of minerals and/or mineraloids.

The Earth's outer solid layer, the lithosphere, is made of rock. In general rocks are of three types, namely, igneous, sedimentary, and metamorphic. The scientific study of rocks is called petrology, and petrology is an essential component of geology.

Rock Classification

Rocks are classified by mineral and chemical composition, by the texture of the constituent particles and by the processes that formed them. These indicators separate rocks into igneous, sedimentary and metamorphic. They are further classified according to particle size. The transformation of one rock type to another is described by the geological model called the rock cycle.

Igneous rocks are formed when molten magma cools and are divided into two main categories: plutonic rock and volcanic. Plutonic or intrusive rocks result when magma cools

Fig 6.1: **The rocky side of a mountain creek near Orosí, Costa Rica**

and crystallizes slowly within the Earth's crust (example granite), while volcanic or extrusive rocks result from magma reaching the surface either as lava or fragmental ejecta (examples pumice and basalt).

Sedimentary rocks are formed by deposition of either clastic sediments, organic matter, or chemical precipitates (evaporites), followed by compaction of the particulate matter and cementation during diagenesis. Sedimentary rocks form at or near the Earth's surface. Mud rocks comprise 65% (mudstone, shale and siltstone); sandstones 20 to 25% and carbonate rocks 10 to 15% (limestone and dolostone).

Metamorphic rocks are formed by subjecting any rock type (including previously-formed metamorphic rock) to different temperature and pressure conditions than those in which the original rock was formed. These temperatures and pressures are always higher than those at the Earth's surface and must be sufficiently high so as to change the original minerals into other mineral types or else into other forms of the same minerals (e.g. by recrystallisation).

The three classes of rocks — the igneous, the sedimentary and the metamorphic — are subdivided into many groups. There are, however, no hard and fast boundaries between allied rocks. By increase or decrease in the proportions of their constituent minerals they pass by every gradation into one another, the distinctive structures also of one kind of rock may often be traced gradually merging into those of another. Hence the definitions adopted in establishing rock nomenclature merely correspond to selected points (more or less arbitrary) in a continuously graduated series.

Impact on Society

Rocks have had a huge impact on the cultural and technological advancement of the human race. Rocks have been used by Homo sapiens and other hominids for more than 2 million years. Lithic techno-logy marks some of the oldest and continuously used technologies. The mining of rocks for their metal ore content has been one of the most important factors of human advancement, which has progressed at different rates in different places in part because of the kind of metals available from the rocks of a region.

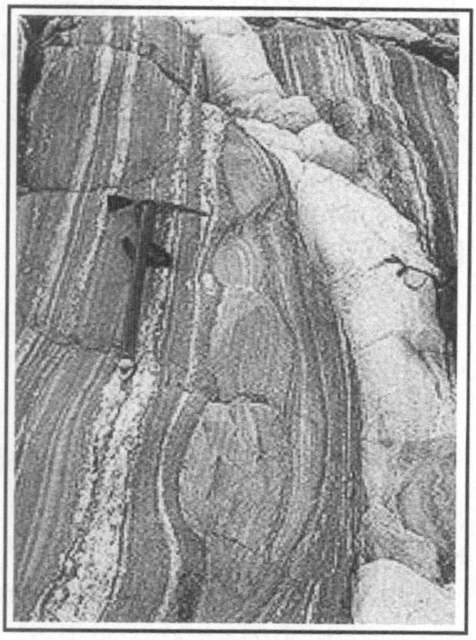

Fig 6.2: **Banded gneiss**

The prehistory and history of civilization is classified into the Stone Age, Bronze Age, and Iron Age. Although the stone age has ended virtually everywhere, rocks continue to be used to construct buildings and infrastructure. When so used, rocks are called dimen-sion stone.

FORMATION

A formation or geological formation is the fundamental unit of lithostratigraphy. A formation consists of a certain number of rock strata that have a comparable lithology, facies or other similar properties. Formations are not defined on the thickness of the rock strata they consist of and the thickness of different formations can therefore vary widely.

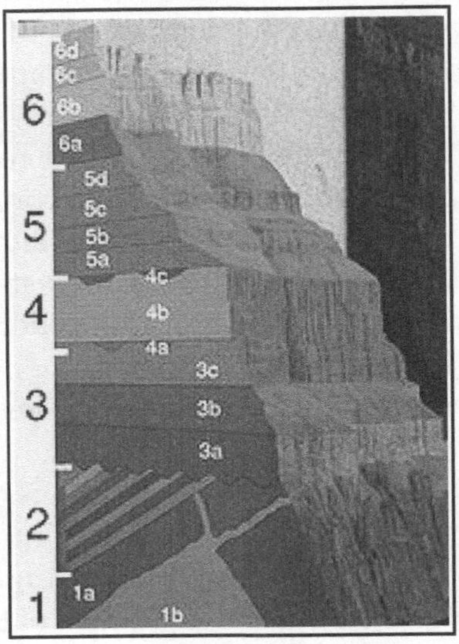

Fig. 6.3: **A geologic cross section of the Grand Canyon**

The concept of formally defined layers or strata is central to the geologic discipline of stratigraphy. A formation can be divided into members and are themselves grouped together in groups.

Usefulness of Formations

Formations allow geologists to correlate geologic strata across wide distances between outcrops and exposures of rock strata.

Formations were initially described to be the essential geologic time markers based on relative ages and the law of superposition. The divisions of the geological time scale were the formations described and put in chronological order by the geologists and stratigraphers of the 17th and 18th centuries.

Modern revision of the geologic sciences has restricted Formations to lithologies, because lithological units are formed by depositional environments, some of which may persist for hundreds of millions of years and will transgress chronostratigraphic intervals or fossil-based methods of correlating rocks. For example, the Hammersley Basin is a Proterozoic sedimentary basin where up to 1200 million years of sedimentation is preserved within the intact sedimentary stratigraphy, with up to 300 million years represented by a single lithological unit of banded iron formation and shale.

Geologic formations are usually sedimentary rock layers, but may also be metamorphic rocks and volcanic flows. Igneous intrusive rocks are generally not divided into Formations.

Defining Lithostratigraphic Formations

Formations are the only formal lithostratigraphic units into which the stratigraphic column everywhere should be divided completely on the basis of lithology.

The establishment varies with the complexity of the geology of a region and the detail needed for geologic mapping and to work out its geologic history.

Formations must be able to be delineated at the scale of geologic mapping practiced in the region. The thickness of formations may range from less than a meter to several thousand meters.

Geologic formations are typically named for the geographic area in which they were first described.

Strictly, formations cannot be defined on any other criteria except primary lithology. However, it is often useful to define Biostratigraphic units based on paleontological criteria, Chronostratigraphic units based on the age of the rocks, and Chemostratigraphic units based on geochemical criteria.

Sequence stratigraphy is a concept which challenges the idea of strict lithostratigraphic units by defining units based on *events* in sedimentary basins such as oceanic regressions and transgressions. These sequences are a combination of chronostratigraphic units, linked by time, and depositional environment linked by the geologic events which occurred at that time, regardless of the grain size of the sediments.

The term "formation" is often used informally to refer to a specific grouping of rocks, such as those encountered within a certain depth range in an oil well.

"Formation" is also used informally to describe the sometimes odd shapes (forms) that rocks acquire through erosional or depositional processes. Some well known cave formations include stalactites and stalagmites.

A geologic formation is said to be abandoned when it is no longer affected by the geologic agent that produced it.

Stratigraphy

Stratigraphy, a branch of geology, studies rock layers and layering (stratification). It is primarily used in the study of sedimentary and layered volcanic rocks. Stratigraphy includes two related subfields: lithologic or lithostratigraphy and biologic stratigraphy or biostratigraphy.

Historical Development

Rock layers were studied since the time of Avicenna (Ibn Sina), a Persian geologist who wrote *The Book of Healing* in 1027. He was the first to outline the law of superposition of strata:

> "It is also possible that the sea may have happened to flow little by little over the land consisting of both plain and mountain, and then have ebbed away from it. . . . It is possible that each time the land was exposed by the ebbing of the sea a layer was left, since we see that some mountains appear to have been piled up layer by layer, and it is therefore likely that the clay from which they were formed was itself at one time arranged in layers. One layer was formed first, then at a different period, a further was formed and piled, upon the first, and so on. Over each layer there spread a substance of different material, which formed a partition between it and the next layer; but when petrification took place

something occurred to the partition which caused it to break up and disintegrate from between the layers (possibly referring to unconformity). . . . As to the beginning of the sea, its clay is either sedimentary or primeval, the latter not being sedimentary. It is probable that the sedimantary clay was formed by the disintegration of the strata of mountains. Such is the formation of mountains."

The theoretical basis for the subject was established by Nicholas Steno who re-introduced the law of superposition and introduced the principle of original horizontality and principle of lateral continuity in a 1669 work on the fossilization of organic remains in layers of sediment.

The first practical large scale application of stratigraphy was by William Smith in the 1790s and early 1800s. Smith, known as the *Father of English Geology*, created the first geologic map of England, and first recognized the significance of strata or rock layering, and the importance of fossil markers for correlating strata. Another influential application of stratigraphy in the early 1800s was a study by Georges Cuvier and Alexandre Brongniart of the geology of the region around Paris.

Lithologic Stratigraphy

Lithostratigraphy, or lithologic stratigraphy, is the most obvious. It deals with the physical lithologic, or rock type, change both vertically in layering or bedding of varying rock type and laterally reflecting changing environments of deposition, known as facies change. Key elements of stratigraphy involve understanding *how* certain geometric relationships between rock layers arise and *what* these geometries mean in terms of depositional environment. One of stratigraphy's basic concepts is codified in the Law of Superposition, which simply states that, in an undeformed stratigraphic sequence, the oldest strata occur at the base of the sequence.

Chemostratigraphy is based on the changes in the relative proportions of trace elements and isotopes within and between lithologic units. Carbon and oxygen isotope ratios vary with time and are used to map subtle changes in the paleoenvironment. This has led to the specialized field of *isotopic stratigraphy*.

Cyclostratigraphy documents the often cyclic changes in the relative proportions of minerals, particularly carbonates, and fossil diversity with time, related to changes in palaeoclimates.

Biostratigraphy

Biostratigraphy or paleontologic stratigraphy is based on fossil evidence in the rock layers. Strata from widespread locations containing the same fossil fauna and flora are correlatable in time. Biologic stratigraphy was based on William Smith's *principle of faunal succession*, which predated, and was one of the first and most powerful lines of evidence for, biological evolution. It provides strong evidence for formation (speciation) of and the extinction of species. The geologic time scale was developed during the 1800s based on the evidence of biologic stratigraphy and faunal succession. This timescale remained a relative scale until the development of radiometric dating, which gave it and the stratigraphy it was based on an absolute time framework, leading to the development of *chronostratigraphy*.

One important development is the Vail curve, which attempts to define a global historical sea-level curve according to inferences from world-wide stratigraphic patterns. Stratigraphy is also commonly used to delineate the nature and extent of hydrocarbon-bearing reservoir rocks, seals and traps in petroleum geology.

Chronostratigraphy

Chronostratigraphy is the branch of stratigraphy that studies the absolute age of rock strata.

Chronostratigraphy is based upon deriving geochronological data for rock units, both directly and by inference, so that a sequence of time relative events of rocks within a region can be derived. In essence, chronostratigraphy seeks to understand the geologic history of rocks and regions.

The ultimate aim of chronostratigraphy is to arrange the sequence of deposition and the time of deposition of all rocks within a geological region, and eventually, the entire geologic record of the Earth.

Magnetostratigraphy

Magnetostratigraphy is a chronostratigraphic technique used to date sedimentary and volcanic sequences. The method works by collecting oriented samples at measured intervals throughout the section. The samples are analyzed to determine their Detrital Remanent Magnetization (DRM), that is, the polarity of Earth's magnetic field at the time a stratum was deposited. This is possible because when very fine-grained magnetic minerals (< 17 micrometres) fall through the water column, they orient themselves with Earth's magnetic field. Upon burial, that orientation is preserved. The minerals, in effect, behave like tiny compasses.

Oriented paleomagnetic core samples are collected in the field; mudstones, siltstones, and very fine-grained sandstones are the preferred lithologies because the magnetic grains are finer and more likely to orient with the ambient field during deposition. If the ancient magnetic field was oriented similar to today's field (North Magnetic Pole near the North Rotational Pole) the strata retain a Normal Polarity. If the data indicate that the North Magnetic Pole was near the South Rotational Pole, the strata exhibit Reversed Polarity.

Results of the individual samples are analysed by removing the Natural Remanent Magnetization (NRM) to reveal the DRM. Following statistical analysis the results are used to generate a local magnetostratigraphic column that can then be compared against the Global Magnetic Polarity Time Scale.

This technique is used to date sequences that generally lack fossils or interbedded igneous rocks. The continuous nature of the sampling means that it is also a powerful technique for the estimation of sediment accumulation rates.

Archaeological Stratigraphy

In the field of archaeology, soil stratigraphy is used to better understand the processes that form and protect archaeological sites. The law of superposition holds true, and this can help date finds or features from each context, as they can be placed in sequence and the dates interpolated. Phases of activity can also often be seen through stratigraphy, especially when a trench or feature is viewed in section (profile). As pits and other features can be dug down into earlier levels, not all material at the same absolute depth is necessarily of the same age, but close attention has to be paid to the archeological layers. The Harris-matrix is a tool to depict complex stratigraphic relations, as they are found, for example, in the contexts of urban archaeology.

7 Mineral

A mineral is a naturally occurring solid formed through geological processes that has a characteristic chemical composition, a highly ordered atomic structure, and specific physical properties. A rock, by comparison, is an aggregate of minerals and/or mineraloids, and need not have a specific chemical composition. Minerals range in composition from pure elements and simple salts to very complex silicates with thousands of known forms The study of minerals is called mineralogy.

Mineral Definition and Classification

To be classified as a true mineral, a substance must be a solid and have a crystalline structure. It must also be a naturally occurring, homogeneous substance with a defined chemical composition. Traditional definitions excluded *organically derived* material. However, the International Mineralogical Association in 1995 adopted a new definition:

> a mineral is an element or chemical compound that is normally crystalline and that has been formed as a result of geological processes.

The modern classifications include an *organic class* - in both the new Dana and the Strunz classification schemes. The *chemical composition* may vary between end members of a mineral system. For example the plagioclase feldspars comprise a continuous series from sodium and silicon-rich albite ($NaAlSi_3O_8$) to calcium and aluminium-rich anorthite ($CaAl_2Si_2O_8$) with four recognized intermediate compositions between. Mineral-like substances that don't strictly meet the definition are sometimes classified as mineraloids. Other natural-occurring substances are nonminerals. Industrial minerals is a market term and refers to commercially valuable mined materials.

A *crystal structure* is the orderly geometric spatial arrangement of atoms in the internal structure of a mineral. There are 14 basic crystal lattice arrangements of atoms in three dimensions, and these are referred to as the 14 "Bravais lattices". Each of these lattices can be classified into one of the six crystal systems, and all crystal structures currently recognized fit in one Bravais lattice and one crystal system. This crystal structure is based on regular internal atomic or ionic arrangement that is often expressed in the geometric form that the crystal takes. Even when the mineral grains are too small to see or are irregularly shaped, the underlying crystal structure is always periodic and can be determined by X-ray diffraction. Chemistry and crystal structure together define a mineral. In fact, two or more minerals may have the same chemical composition, but differ in crystal structure (these are known as *polymorphs*). For example, pyrite and marcasite are both iron sulfide, but their arrangement of atoms differs. Similarly, some minerals have different chemical compositions, but the same crystal structure: for example, halite (made from sodium and chlorine), galena (made from lead and sulfur) and periclase (made from magnesium and oxygen) all share the same cubic crystal structure.

Crystal structure greatly influences a mineral's physical properties. For example, though diamond and graphite have the same composition (both are pure carbon), graphite is very soft, while diamond is the hardest of all known minerals. This happens because the carbon atoms in graphite are arranged into sheets which can slide easily past each other, while the carbon atoms in diamond form a strong, interlocking three-dimensional network.

There are currently more than 4,000 known minerals, according to the International Mineralogical Association, which is responsible for the approval of and naming of new mineral species found in nature. Of these, perhaps 100 can be called "common", 50 are "occasional", and the rest are "rare" to "extremely rare".

Differences Between Minerals and Rocks

A mineral is a naturally occurring solid with a definite chemical composition and a specific crystalline structure. A rock is an aggregate of one or more minerals. (A rock may also include organic remains and mineraloids.) Some rocks are predominantly composed of just one mineral. For example, limestone is a sedimentary rock composed almost entirely of the mineral calcite. Other rocks contain many minerals, and the specific minerals in a rock can vary widely. Some minerals, like quartz, mica or feldspar are common, while others have been found in only four or five locations worldwide. The vast majority of the rocks of the Earth's crust consist of quartz, feldspar, mica, chlorite, kaolin, calcite, epidote, olivine, augite, hornblende, magnetite, hematite, limonite and a few other minerals. Over half of the mineral species known are so rare that they have only been found in a handful of samples, and many are known from only one or two small grains.

Commercially valuable minerals and rocks are referred to as industrial minerals. Rocks from which minerals are mined for economic purposes are referred to as ores (the rocks and minerals that remain, after the desired mineral has been separated from the ore, are referred to as tailings).

Mineral Composition of Rocks

A main determining factor in the formation of minerals in a rock mass is the chemical composition of the mass, for a certain mineral can be formed only when the necessary elements are present in the rock. Calcite is most common in limestones, as these consist essentially of calcium carbonate; quartz is common in sandstones and in certain igneous rocks which contain a high percentage of silica.

Other factors are of equal importance in determining the natural association or paragenesis of rock-forming minerals, principally the mode of origin of the rock and the stages through which it has passed in attaining its present condition. Two rock masses may have very much the same bulk composition and yet consist of entirely different assemblages of minerals. The tendency is always for those compounds to be formed which are stable under the conditions under which the rock mass originated. A granite arises by the consolidation of a molten magma at high temperatures and great pressures and its component minerals are those stable under such conditions. Exposed to moisture, carbonic acid and other subaerial agents at the ordinary temperatures of the Earth's surface, some of these original minerals, such as quartz and white mica are relatively stable and remain unaffected; others weather or decay and are replaced by new combinations. The feldspar passes into kaolinite, muscovite and quartz, and any mafic minerals such as pyroxenes, amphiboles or biotite have been present they are often altered to chlorite, epidote, rutile and other substances. These changes are accompanied by disintegration, and the rock falls into a loose, incoherent, earthy mass which may be regarded as a sand or soil. The materials thus formed may be washed away and deposited as sandstone or siltstone. The structure of the original rock is now replaced by a new one; the mineralogical constitution is profoundly altered; but the bulk chemical composition may not be very different. The sedimentary rock may again undergo metamorphism. If penetrated by igneous rocks it may be recrystallized or, if subjected to enormous pressures with heat and movement during mountain building, it may be converted into a gneiss not very different in mineralogical composition though radically different in structure to the granite which was its original state.

Physical Properties of Minerals

Classifying minerals can range from simple to very difficult. A mineral can be identified by several physical properties, some of them being sufficient for full identification without equivocation. In other cases, minerals can only be classified by more complex chemical or X-ray diffraction analysis; these methods, however, can be costly and time-consuming.

Physical properties commonly used are:

- *Crystal structure and habit*: See the above discussion of crystal structure. A mineral may show good crystal habit or form, or it may be massive, granular or compact with only microscopically visible crystals.
- *Hardness*: the physical hardness of a mineral is usually measured according to the Mohs scale. This scale is relative and goes from 1 to 10. Minerals with a given Mohs hardness can scratch the surface of any mineral that has a lower hardness than itself.

Mohs hardness scale:

1. Talc $Mg_3Si_4O_{10}(OH)_2$
2. Gypsum $CaSO_4$ $2H_2O$
3. Calcite $CaCO_3$
4. Fluorite CaF_2
5. Apatite $Ca_5(PO_4)_3(OH, Cl, F)$
6. Orthoclase $KAlSi_3O_8$
7. Quartz SiO_2
8. Topaz $Al_2SiO_4(OH, F)_2$

9. Corundum Al_2O_3
10. Diamond C (pure carbon)

- *Luster* indicates the way a mineral's surface interacts with light and can range from dull to glassy (vitreous).
  - Metallic-high reflectivity like metal: galena and pyrite.
  - Sub-metallic-slightly less than metallic reflectivity: magnetite.
  - Non-metallic lusters:
    - ❑ *Adamantine* - brilliant, the luster of diamond also cerussite and anglesite.
    - ❑ *Vitreous* - the luster of a broken glass: quartz.
    - ❑ *Pearly* - iridescent and pearl-like: talc and apophyllite
    - ❑ *Resinous* - the luster of resin: sphalerite and sulfur
    - ❑ *Silky* - a soft light shown by fibrous materials: gypsum and chrysotile.
    - ❑ Dull/earthy - shown by finely crystallized minerals: the kidney ore variety of hematite.
- *Colour* indicates the appearance of the mineral in reflected light or transmitted light for translucent minerals (i.e. what it looks like to the naked eye).
  - *Iridescence* - the *play of colors* due to surface or internal interference. Labradorite exhibits internal iridescence whereas hematite and sphalerite often show the surface effect.
- *Streak* refers to the color of the powder a mineral leaves after rubbing it on an unglazed porcelain *streak plate*. Note that this is not always the same color as the original mineral.

- *Cleavage* describes the way a mineral may split apart along various planes. In thin sections, cleavage is visible as thin parallel lines across a mineral.
- *Fracture* describes how a mineral breaks when broken contrary to its natural cleavage planes.
  - *Chonchoidal fracture* is a smooth curved fracture with concentric ridges of the type shown by glass.
  - *Hackley* is jagged fracture with sharp edges.
  - *Fibrous*
  - *Irregular*
- *Specific gravity* relates the mineral mass to the mass of an equal volume of water, namely the density of the material. While most minerals, including all the common rock-forming minerals, have a specific gravity of 2.5-3.5, a few are noticeably more or less dense, e.g. several sulfide minerals have high specific gravity compared to the common rock-forming minerals.
- *Other properties:* fluorescence (response to ultraviolet light), magnetism, radioactivity, tenacity (response to mechanical induced changes of shape or form), piezoelectricity and reactivity to dilute acids.

Chemical Properties of Minerals

Minerals may be classified according to chemical composition. They are here categorized by anion group. The list below is in approximate order of their abundance in the Earth's crust. The list follows the Dana classification system which closely parallels the Strunz classification.

Silicate Class

The largest group of minerals by far are the silicates (most rocks are =95% silicates), which are composed largely of silicon and oxygen, with the addition of ions such as aluminium, magnesium, iron, and calcium. Some important rock-forming silicates include the feldspars, quartz, olivines, pyroxenes, amphiboles, garnets, and micas.

Carbonate Class

The carbonate minerals consist of those minerals containing the anion $(CO_3)^{2-}$ and include calcite and aragonite (both calcium carbonate), dolomite (magnesium/calcium carbonate) and siderite (iron carbonate). Carbonates are commonly deposited in marine settings when the shells of dead planktonic life settle and accumulate on the sea floor. Carbonates are also found in evaporitic settings (e.g. the Great Salt Lake, Utah) and also in karst regions, where the dissolution and reprecipitation of carbonates leads to the formation of caves, stalactites and stalagmites. The carbonate class also includes the nitrate and borate minerals.

Sulfate Class

Sulfates all contain the sulfate anion, SO_4^{2-}. Sulfates commonly form in evaporitic settings where highly saline waters slowly evaporate, allowing the formation of both sulfates and halides at the water-sediment interface. Sulfates also occur in hydrothermal vein systems as gangue minerals along with sulfide ore minerals. Another occurrence is as secondary oxidation products of original sulfide minerals. Common sulfates include anhydrite (calcium sulfate), celestine (strontium sulfate), barite (barium sulfate), and gypsum (hydrated calcium sulfate). The sulfate class also includes the chromate, molybdate, selenate, sulfite, tellurate, and tungstate minerals.

Halide Class

The halides are the group of minerals forming the natural salts and include fluorite (calcium fluoride), halite (sodium chloride), sylvite (potassium chloride), and sal ammoniac (ammonium chloride). Halides, like sulfates, are commonly found in evaporitic settings such as playa lakes and landlocked seas such as the Dead Sea and Great Salt Lake. The halide class includes the fluoride, chloride, bromide and iodide minerals.

Oxide Class

Oxides are extremely important in mining as they form many of the ores from which valuable metals can be extracted. They also carry the best record of changes in the Earth's magnetic field. They commonly occur as precipitates close to the Earth's surface, oxidation products of other minerals in the near surface weathering zone, and as accessory minerals in igneous rocks of the crust and mantle. Common oxides include hematite (iron oxide), magnetite (iron oxide), chromite (iron chromium oxide), spinel (magnesium aluminium oxide - a common component of the mantle), ilmenite (iron titanium oxide), rutile (titanium dioxide), and ice (hydrogen oxide). The oxide class includes the oxide and the hydroxide minerals.

Sulfide Class

Many sulfide minerals are economically important as metal ores. Common sulfides include pyrite (iron sulfide - commonly known as *fools' gold*), chalcopyrite (copper iron sulfide), pentlandite (nickel iron sulfide), and galena (lead sulfide). The sulfide class also includes the selenides, the tellurides, the arsenides, the antimonides, the bismuthinides, and the sulfosalts (sulfur and a second anion such as arsenic).

Phosphate Class

The phosphate mineral group actually includes any mineral with a tetrahedral unit AO_4 where A can be phosphorus, antimony, arsenic or vanadium. By far the most common phosphate is apatite which is an important biological mineral found in teeth and bones of many animals. The phosphate class includes the phosphate, arsenate, vanadate, and antimonate minerals.

The elemental group includes metals and intermetallic elements (gold, silver, copper), semi-metals and non-metals (antimony, bismuth, graphite, sulfur). This group also includes natural alloys, such as electrum (a natural alloy of gold and silver), phosphides, silicides, nitrides and carbides (which are usually only found naturally in a few rare meteorites).

Organic Class

The organic mineral class includes biogenic substances in which geological processes have been a part of the genesis or origin of the existing compound. Minerals of the organic class include various oxalates, mellitates, citrates, cyanates, acetates, formates, hydrocarbons and other miscellaneous species. Examples include whewellite, moolooite, mellite, fichtelite, carpathite, evenkite and abelsonite.

8 Oxides and Hydroxides

An oxide is a chemical compound contaning at least one oxygen atom as well as at least one other element. Most of the Earth's crust consists of oxides. Oxides result when elements are oxidized by oxygen in air. Combustion of hydrocarbons affords the two principal oxides of carbon, carbon monoxide and carbon dioxide. Even materials that are considered to be pure elements often contain a coating of oxides. For example, aluminium foil has a thin skin of Al_2O_3 that protects the foil from further corrosion.

Virtually all elements burn in an atmosphere of oxygen. In the presence of water and oxygen (or simply air), some elements - lithium, sodium, potassium, rubidium, caesium, strontium and barium - react rapidly, even dangerously, to give the hydroxides. In part for this reason, alkali and alkaline earth metals are not found in nature in their metallic, i.e., native, form. Caesium is so reactive with oxygen that it is used as a getter in vacuum tubes, and solutions of potassium and sodium, so called NaK are used to deoxygenate and dehydrate some organic solvents. The surface of most metals consist of oxides and hydroxides in the presence of air. A well known example is aluminium foil, which is coated with a thin film of aluminium oxide that passivates the metal, slowing further corrosion. The aluminium oxide layer can be built to greater thickness by the process of electrolytic anodising. Although solid magnesium and aluminium react slowly with oxygen at STP, they, like most metals, will burn in air, generating very high temperatures. As a consequence, finely grained powders of most metals can be dangerously explosive in air.

In dry oxygen, iron readily forms iron(II) oxide, but the formation of the hydrated ferric oxides, $Fe_2O_{3-2x}(OH)_x$, that mainly comprise rust, typically requires oxygen *and* water. The production of free oxygen by photosynthetic bacteria some 3.5 billion years ago precipitated iron out of solution in the oceans as Fe_2O_3 in the economically-important iron ore hematite.

Due to its electronegativity, oxygen forms chemical bonds with almost all elements to give the corresponding oxides. So-called noble metals (common examples: gold, platinum) resist direct chemical combination with oxygen, and substances like gold(III) oxide must be generated by indirect routes.

Insolubility in Water

The oxide ion, O^{2-}, is the conjugate base of the hydroxide ion, OH^-, and is encountered in ionic solid such as calcium oxide. O^{2-} is unstable in aqueous solution - its affinity for H^+ is so great ($pK_b \sim -22$) that it abstracts a proton from a solvent H_2O molecule:

$$O^{2-} + H_2O = 2\ OH^-$$

Although many anions are stable in aqueous solution, ionic oxides are not. For example, sodium chloride dissolves readily in water to give a solution containing the constituent ions, Na^+ and Cl^-. Oxides do not behave like this. If an ionic oxide dissolves, the O^{2-} ions become protonated. Although calcium oxide, CaO, is said to "dissolve" in water, the products include hydroxide:

$CaO + H_2O = Ca^{2+} + 2\ OH^-$

In fact, no monoatomic dianion is known to dissolve in water - all are so basic that they undergo hydrolysis. Concentrations of oxide ion in water are too low to be detectable with current technology.

Authentic soluble oxides do exist, but they release oxyanions, not O^{2-}. Well known soluble salts of oxyanions include sodium sulfate (Na_2SO_4), potassium permanganate ($KMnO_4$), and sodium nitrate ($NaNO_3$).

Nomenclature

In the 18th century, oxides were named *calxes* or *calces* after the calcination process used to produce oxides. *Calx* was later replaced by *oxyd.*

Oxides are usually named after the number of oxygen atoms in the oxide. Oxides containing only one oxygen are called oxides or monoxides, those containing two oxygen atoms are *dioxides*, three oxygen atoms makes it a *trioxide*, four oxygen atoms are tetroxides, and so on following the Greek numerical prefixes. In the older literature and continuing in industry, oxides are named by contracting the element name with "a." Hence alumina, magnesia, chromia, are, respectively, Al_2O_3, MgO, Cr_2O_3.

Two other types of oxide are peroxide, O_2^{2-}, and superoxide, O_2^-. In such species, oxygen is assigned higher oxidation states than oxide.

Types of Oxides

Oxides of more electropositive elements tend to be basic. They are called *basic anhydrides;* adding water, they may form basic hydroxides. For example, sodium oxide is basic; when hydrated, it forms sodium hydroxide.

Oxides of more electronegative elements tend to be acidic. They are called *acid anhydrides;* adding water, they form oxoacids. For example, dichlorine heptoxide is acid; perchloric acid is a more hydrated form.

Some oxides can act as both acid and base at different times. They are amphoteric. An example is aluminium oxide. Some oxides do not show behavior as either acid or base.

The oxides of the chemical elements in their highest oxidation state are predictable and the chemical formula can be derived from the number of valence electrons for that element. Even the chemical formula of O_4, tetraoxygen, is predictable as a group 16 element. One exception is copper for which the highest oxidation state oxide is copper(II) oxide and not copper(I) oxide. Another exception is fluoride that does not exist as expected as F_2O_7 but as OF_2. Since F is more electronegative than O, OF_2 does not represent an oxide of fluorine, but instead represents a fluoride of oxygen. Phosphorus pentoxide, the third exception is not properly represented by the chemical formula P_2O_5 but by P_4O_{10}.

List of all known oxides sorted by oxidation state

- Element in -1 oxidation state
  - Oxygen difluoride (OF_2)
- Element in multiple oxidation states
  - Antimony tetroxide (Sb_2O_4)
  - Cobalt (II, III) oxide (Co_3O_4)
  - Iron (II, III) oxide (Fe_3O_4)
  - Lead tetroxide (Pb_3O_4)
  - Manganese (II,III) oxide (Mn_3O_4)
  - Silver (I, III) oxide (AgO)
- Element in +1 oxidation state
  - Copper(I) oxide (Cu_2O)
  - Dicarbon monoxide (C_2O)
  - Dichlorine monoxide (Cl_2O)
  - Lithium oxide (Li_2O)

- Potassium oxide (K_2O)
- Rubidium oxide (Rb_2O)
- Silver(I) oxide (Ag_2O)
- Thallium oxide (Tl_2O)
- Sodium oxide (Na_2O)
- Water (hydrogen oxide) (H_2O)

• Element in +2 oxidation state

- Aluminium monoxide (AlO)
- Barium oxide (BaO)
- Beryllium oxide (BeO)
- Cadmium oxide (CdO)
- Calcium oxide (CaO)
- Carbon monoxide (CO)
- Cobalt (II) oxide (CoO)
- Copper (II) oxide (CuO)
- Iron (II) oxide (FeO)
- Lead (II) oxide (PbO)
- Magnesium oxide (MgO)
- Mercury (II) oxide (HgO)
- Nickel (II) oxide (NiO)
- Nitrogen oxide (NO)
- Palladium (II) oxide (PdO)
- Strontium oxide (SrO)
- Sulphur monoxide (SO)
- Tin (II) oxide (SnO)
- Titanium (II) oxide (TiO)

- Vanadium (II) oxide (VO)
- Zinc oxide (ZnO)

- Element in +3 oxidation state
  - Aluminium oxide (Al_2O_3)
  - Antimony trioxide (Sb_2O_3)
  - Arsenic trioxide (As_2O_3)
  - Bismuth trioxide (Bi_2O_3)
  - Boron oxide (B_2O_3)
  - Chromium (III) oxide (Cr_2O_3)
  - Dinitrogen trioxide (N_2O_3)
  - Erbium (III) oxide (Er_2O_3)
  - Gadolinium (III) oxide (Gd_2O_3)
  - Gallium (III) oxide (Ga_2O_3)
  - Holmium (III) oxide (Ho_2O_3)
  - Indium (III) oxide (In_2O_3)
  - Iron (III) oxide (Fe_2O_3)
  - Lanthanum (III) oxide (La_2O_3)
  - Lutetium (III) oxide (Lu_2O_3)
  - Nickel (III) oxide (Ni_2O_3)
  - Phosphorus trioxide (P_4O_6)
  - Promethium (III) oxide (Pm_2O_3)
  - Rhodium (III) oxide (Rh_2O_3)
  - Samarium (III) oxide (Sm_2O_3)
  - Scandium (III) oxide (Sc_2O_3)
  - Terbium (III) oxide (Tb_2O_3)
  - Thallium (III) oxide (Tl_2O_3)

- Thulium (III) oxide (Tm_2O_3)
- Titanium (III) oxide (Ti_2O_3)
- Tungsten (III) oxide (W_2O_3)
- Vanadium (III) oxide (V_2O_3)
- Ytterbium (III) oxide (Yb_2O_3)
- Yttrium (III) oxide (Y_2O_3)

• Element in +4 oxidation state

- Carbon dioxide (CO_2)
- Carbon trioxide (CO_3)
- Cerium (IV) oxide (CeO_2)
- Chlorine dioxide (ClO_2)
- Chromium (IV) oxide (CrO_2)
- Dinitrogen tetroxide (N_2O_4)
- Germanium dioxide (GeO_2)
- Hafnium (IV) oxide (HfO_2)
- Lead (IV) oxide (PbO_2)
- Manganese (IV) oxide (MnO_2)
- Nitrogen dioxide (NO_2)
- Plutonium dioxide (PuO_2)
- Ruthenium (IV) oxide (RuO_2)
- Selenium dioxide (SeO_2)
- Silicon dioxide (SiO_2)
- Sulfur dioxide (SO_2)
- Tellurium dioxide (TeO_2)
- Thorium dioxide (ThO_2)
- Tin dioxide (SnO_2)

- Titanium dioxide (TiO_2)
- Tungsten (IV) oxide (WO_2)
- Uranium dioxide (UO_2)
- Vanadium (IV) oxide (VO_2)
- Zirconium dioxide (ZrO_2)

• Element in +5 oxidation state
- Antimony pentoxide (Sb_2O_5)
- Arsenic pentoxide (As_2O_5)
- Dinitrogen pentoxide (N_2O_5)
- Niobium pentoxide
- Phosphorus pentoxide (P_2O_5)
- Tantalum pentoxide (Ta_2O_5)
- Vanadium(V) oxide (V_2O_5)

• Element in +6 oxidation state
- Chromium trioxide (CrO_3)
- Molybdenum (VI) oxide (MoO_3)
- Rhenium trioxide (ReO_3)
- Selenium trioxide (SeO_3)
- Sulphur trioxide (SO_3)
- Tellurium trioxide (TeO_3)
- Tungsten trioxide (WO_3)
- Uranium trioxide (UO_3)
- Xenon trioxide (XeO_3)

• Element in +7 oxidation state
- Dichlorine heptoxide (Cl_2O_7)
- Manganese (VII) oxide (Mn_2O_7)
- Rhenium (VII) oxide (Re_2O_7)

- Technetium (VII) oxide

- Element in +8 oxidation state
  - Osmium tetroxide (OsO_4)
  - Ruthenium tetroxide (RuO_4)
  - Xenon tetroxide (XeO_4)

Properties of Oxides and Hydroxides

The two most important properties of oxides and hydroxides are:

- their structure and bonding
- their acid-base properties.

We shall consider the oxides of Group 1, Group 2, B and Al from group 3, C, Si and Pb from group 4, P, S, Cl, and the first *d* block.

Definitions

- *Ionic* Oxides and hydroxides are ionic if they have high melting and boiling temperatures and a crystal structure which is composed of separate ions.
- *Covalent* Oxides and oxo-acids are molecular if they have low melting and boiling temperatures and they are made up of discrete molecules. A giant covalent compound has a high melting and boiling temperature with a continuous structure.
- *Layer lattice.* Many hydroxides have a layer lattice structure with metal ions sandwiched between layers of hydroxide ions. These are ionic/covalent in character. On heating to a high temperature, they give ionic oxides.
- *Basic.* Metals with electronegativity less than 1.3 form basic oxides and hydroxides. These oxides and hydroxides dissolve easily in dilute acids. If they are soluble in water they give rise to alkaline solutions.

- *Acidic.* Non-metals with electronegativity greater than 2.0 form acidic oxides. If these are soluble in water they give oxo-acids which are acidic. The giant covalent oxides B_2O_3 and SiO_2 do not dissolve in water but are classified as acidic as they react with fused NaOH; in addition their oxo-acids, prepared by hydrolysis of the chlorides, dissolve in aqueous NaOH.
- *Amphoteric.* Metals with electronegativity between 1.3 and 2.0 should form amphoteric oxides but many of them have very insoluble hydroxides which dissolve only slightly in aqueous NaOH. These oxides will be considered to be amphoteric only if their hydroxides dissolve readily in aqueous NaOH; otherwise they will be considered basic.
- *Formula.* If the hydroxy compound is basic or amphoteric it will be written in the form $M(OH)_x$, even though many tri-positive metals form hydroxides with the formula $MO.OH._xH_2O$.. If the hydroxy compound is acidic, it will be written in the form H_xXO_y.

Group 1 and Group 2

All the Group 1 oxides are ionic and alkaline:

$$M_2O\ (s) + H_2\ O(l) \rightarrow 2MOH(aq)$$

The same is true for Group 2 oxides:

$$MO(s) + H_2O(l) \rightarrow M(OH)_2\ (aq)$$

although beryllium and magnesium hydroxides are nearly insoluble (solubility of hydroxides increases down the group).

Beryllium oxide dissolves in both dilute acid and alkali and so is amphoteric (compare with aluminium):

$$BeO(s) + 2H^+(aq) \rightarrow Be^{2+}(aq) + H_2O(l)$$

$$BeO(s) + 2OH^- + H_2O(l) \rightarrow Be(OH)_4^{2-}(aq)$$

Group 3

Boron oxide is giant covalent and is acidic. It is insoluble in water but dissolves in fused NaOH. The weak acid, H_3BO_3, made by hydrolysis of BCl_3, dissolves in aqueous NaOH (compare silicon).

$$B_2O_3(s) + NaOH(l) \rightarrow 2NaBO_2(l) + H_2O(g)$$

$$H_3BO_3(s) + OH^-(aq) \rightarrow B(OH)_4{}^-(aq)$$

Aluminium oxide is ionic and the hydroxide forms a layer lattice. Its oxide shows similar behaviour to that of boron (substitute Al for B in the equations above), and as it also dissolves in acids, it is amphoteric.

$$Al_2O_3(s) + 6H^+(aq) \rightarrow 2Al^{3+}(aq) + 3H_2O(l)$$

$$Al(OH)_3(s) + 3H^+(aq) \rightarrow Al^{3+}(aq) + 3H_2O(l)$$

The resulting aluminium ion, $Al(H_2O)_6{}^{3+}(aq)$, has a high charge density and acts as a weak acid with pK_a 4.8, the same as ethanoic acid.

$$[Al(H_2O)_6{}^{3+}(aq) + H_2O(l) \rightarrow [Al(H_2O)_5OH]^{2+} + H_3O^+(aq)$$

On the addition of aqueous NaOH, the equilibrium moves to the right until $Al(OH)_3(s)$ is precipitated; this easily reacts with excess NaOH to give the aluminate ion, $Al(OH)_4{}^-$ (aq).

Group 4

Carbon monoxide is molecular and is considered to be neutral. Formally it is the anhydride of methanoic acid HCOOH, but it does not react with water although the reverse process, the dehydration of methanoic acid with concentrated sulphuric acid, is a standard method for preparing carbon monoxide.

Carbon dioxide is molecular and dissolves in water to give carbonic acid, which is a weak acid with pK_1 6.4:

$$CO_2(g) + H_2O(l) \rightarrow H_2CO_3(aq)$$

Only about 1 molecule in 600 of $CO_2(g)$ is converted into the acid and most of it remains in the form $CO_2(aq)$.

Silicon dioxide is giant covalent , containing single Si-O bonds. A molecular structure would contain Si=O bonds; these would be much weaker than two of the Si-O single bonds found in the silica lattice. SiO2 is insoluble in water but reacts with fused NaOH:

$$SiO_2(s) + 2NaOH(l) \rightarrow Na_2SiO_3(l) + H_2O(g)$$

The very weak acid, $H_4SiO_4(s)$, prepared by the hydrolysis of the chloride, also dissolves in aqueous NaOH.

$$H_4SiO_4(s) + 2OH^-(aq) \rightarrow SiO_3{}^{2-}(aq) + 3H_2O(l)$$

Lead monoxide is an ionic solid whose colour varies from yellow to orange. It is red when hot and turns back to yellow on cooling. It disolves in dilute nitric acid (the sulphate and chloride are insoluble in cold water) and in concentrated aqeuous sodium hydroxide.

$$PbO(s) + 2H^+(aq) \rightarrow Pb^{2+}(aq) + H_2O(l)$$

$$PbO(s) + H_2O(l) + OH^-(aq) \rightarrow Pb(OH)_3{}^-(aq)$$

Addition of aqueous NaOH to a solution of a lead salt gives a white precipitate usually written as $Pb(OH)_2(s)$ (but possibly $2PbO.H_2O$)which reacts with excess NaOH to give $Pb(OH)_3^-(aq)$, plumbate (II).

Lead dioxide, $PbO_2(s)$ is an ionic solid which is brown. It decomposes to PbO and oxygen on heating. At low temperature it dissolves in concentrated HCl to give PbCl4 and the complex ion $PbCl_6^{2-}$. On warming, these substances give off chlorine gas. $PbO_2(s)$ also dissolves in concentrated NaOH.

$$PbO_2(s) + 2OH^- (aq) + 2H_2O (l) \rightarrow pb(OH)_6{}^{2-}(aq)$$

Lead (IV) hydroxide is unknown; $PbO_2(s)$ is precipitated instead.

Phosphorus, Sulphur and Chlorine

All form molecular oxides which are acidic.

Phosphorus (III) oxide and phosphorus(V) oxide have molecular formulae P_4O_6(s) and P_4O_{10}(s) but are often seen written as P_2O_3(s) and P_2O_5(s). They react with water to give the weak acid H_3PO_3 with pK_1 2.0, and the slightly stronger acid, H_3PO_4 with pK_1 1.49. This illustrates the general trend that the oxide of the higher oxidation state is the more acidic.

$$P_2O_3(s) + 2H_2O(l) \rightarrow 2H_3PO_3(aq)$$

$$P_2O_5(s) + 2H_2O(l) \rightarrow 2H_3PO_4(aq)$$

Sulphur dioxide and sulphur trioxide dissolve in water to give the weak acid, H_2SO_3 with pK_1 1.81), and the very strong acid, H_2SO_4. Note that even in this case the second ionisation is small enough for it to have very little effect on the pH; see the 'pH of sulphuric acid' page.

Chlorine forms several oxides of which Cl_2O and Cl_2O_7 are the most important. They react with water to form the weak acid HClO with pK_a 7.53), and the very strong acid, $HClO_4$.

$$Cl_2O(g) + 2H_2O(l) \rightarrow 2HClO(aq)$$

$$Cl_2O_7(l) + 2H_2O(l) \rightarrow 2HClO_4(aq)$$

3d Block Metals

Scandium oxide is very similar to aluminium oxide.

The most important oxide of titanium is TiO_2 which is acidic.

Vanadium has four principal oxidation states. VO and V_2O_3 are both ionic and basic and dissolve in acids to give the lilac $[V(H_2O)_6]^{2+}$ ion and the blue-green $[V(H_2O)_6]^{3+}$ ion. Both of these ions are readily oxidised to V(IV) or V(V). VO_2 is ionic and amphoteric. It dissolves in acids to give the blue VO^{2+} ion (probably $[VO(H_2O)_5]^{2+}$) and in concentrated alkali to give the brown VO_4^{4-} ion. V_2O_5 has a borderline ionic/

covalent structure. It reversibly loses oxygen on heating, and this may be why it is such an efficient catalyst in the Contact Process for the conversion of sulphur dioxide to sulphur trioxide. It is amphoteric.

$V_2O_5(s) + 2H^+(aq) \rightarrow 2VO_2{}^+(aq) + H_2O(l)$ yellow solution

$V_2O_5(s) + 6OH^-(aq) \rightarrow 2VO_4{}^{3-}(aq) + 3H_2O(l)$ colourless solution

The two main oxides of chromium are the green Cr_2O_3 which is ionic, and the red CrO_3 which is covalent. The former is amphoteric as the green-grey precipitate of $Cr(OH)_3$ dissolves in acids to give the purple $[Cr(OH)_6]^{3+}$ ion, and in alkali to give a green solution which may possibly contain the $Cr(OH)_4{}^-$ ion. With ammonia solution, the green-grey precipitate dissolves to give a pink-purple $[Cr(NH_3)_6]^{3+}$ complex ion.

$[Cr(H_2O)_6]^{3+}(aq) + 3OH^-(aq) \rightarrow Cr(OH)_3(s) + 6H_2O(l)$

This is a deprotonation or acid-base reaction.

$Cr(OH)_3(s) + 6NH_3(aq) \rightarrow [Cr(NH_3)_6]^{3+}(aq) + 3\,OH^-(aq)$

This is a ligand exchange reaction.

Chromium trioxide is acidic and dissolves in water to give the fairly strong acid, H_2CrO_4 with pK_a 0.74. On heating it loses oxygen to give Cr_2O_3. When treated with alkali the solution contains the yellow $CrO_4{}^{2-}(aq)$ ion. This reacts with strong acids to give an orange solution containing the dichromate (VI) ion:

$CrO_3(s) + H_2O(l) \rightarrow H_2CrO_4(aq)$

$2CrO_4{}^{2-}(aq) + 2H^+(aq) \rightarrow Cr_2O_7{}^{2-}(aq) + H_2O(l)$

Manganese(II) compounds contain the very pale pink d5 ion which is resistant to oxidation. If alkali is added to a solution containing the $Mn^{2+}(aq)$ ion, the off white precipitate of $Mn(OH)_2$ is readily oxidised by air to brown hydrated Mn_2O_3 and then possibly to $2MnO_2.H_2O$. Reduction of the purple MnO_4^- ion in alkaline solution (for example by alkenes) yields brown hydrated MnO_2.

Iron (II) oxide FeO is difficult to prepare as it rapidly oxidises in air to give Fe_2O_3, which contains the d^5 ion. Both of these oxides are basic. Freshly precipitated $Fe(OH)_2$ is green but it rapidly darkens on exposure to air and finally forms brown $Fe(OH)_3$.

Pink cobalt compounds contain the $[Co(H_2O)_6]^{2+}$ ion. On the addition of NaOH a precipitate of $Co(OH)_2$ is formed which starts off blue but turns pink on warming. The blue form may have four, rather than six, OH^- groups round each cobalt ion in the solid state. With ammonia $Co(OH)_2$ dissolves giving a complex which undergoes air oxidation to give the brown $[Co(NH_3)_6]^{3+}$(aq) ion.

Solutions of nickel(II) containing the green $[Ni(H_2O)_6]^{2+}$(aq) ion give, with NaOH, a pale green precipitate of $Ni(OH)_2$ which dissolves in ammonia to give the purple $[Ni(NH_3)_6]^{2+}$(aq) ion.

When copper(I) oxide Cu_2O(s) is treated with aqueous sulphuric acid, the Cu^+(aq) ion initially formed disproportionates to a brown precipitate of Cu(s) and the blue Cu^{2+}(aq) ion, The hydrated $[Cu(H_2O)_6]^{2+}$(aq) ion has four water molecules more firmly bound than the other two, and it is these water molecules which are substituted when pale blue $Cu(OH)_2$(s) dissolves in aqueous ammonia to give the deep blue $[Cu(NH_3)_4(H_2O)_2]^{2+}$(aq) ion.

Zinc oxide goes yellow on heating and returns to white again on cooling. This is result of movement of ions within the crystal lattice which is reversible. It is amphoteric, giving the $[Zn(H_2O)_4]^{2+}$(aq) ion with acids and the $Zn(OH)_3^-$(aq) and $Zn(OH)_4^{2-}$(aq) ions when treated with NaOH. It also reacts with ammonia (contrast $Al(OH)_3$(s), which does not) to give the $[Zn(NH_3)_4]^{2+}$(aq) ion.

A high redox potential equals to well-aerated environmental conditions and a low redox potential equals to saturated environmental conditions. Saturated soils become depleted of oxygen, because this is rapidly consumed by aerobic organisms and cannot be replenished by diffusion quickly. Then, anaerobic and facultative organisms continue the decomposition process. In the absence of oxygen, other electron acceptors begin to function, depending on their tendency to accept electrons. When flooding occurs the reduction of the remaining oxygen will take place first, followed by the reduction of nitrate, then manganese, iron, sulphate, and carbon dioxide (The reduction of oxygen occurs by the O_2 consumption of aerobic organisms, NO_3^- serves as a biochemical electron acceptor involving N-organisms that ultimately excrete reduced N, the reduction of Mn can be initiated in presence of NO_3^-, whereas the reduction of Fe cannot be initiated in presence of NO_3^-, and sulfate reducing bacteria are involved to reduce SO_4^{2-}.

The oxides and hydroxides present in soils reflect the pedoenvironmental conditions of soil formation. The parent material, temperature, moisture, organic material, pH, and Eh control the formation of different types of oxides and hydroxides. Because the oxides and hydroxides, particularly of iron and manganese, show different colors they can be used as an indicator for processes of pedogenesis. It should be stressed that there are continual modifications of pedogenic processes acting in soils. Therefore, the soil properties which can be observed in the field, e.g. soil color expressed by the presence or absence of oxides and hydroxides and the distribution of them, also changes. Thus, it is also important to relate the oxides and hydroxides to contemporary processes of pedogenesis or soil formation associated to some previous periods.

Characteristics of oxides and hydroxides in soils are a relative high cation exchange capacity (CEC), which is due to the dissociation of protons from -OH and $-OH_2$ groups of the hydroxides. This is true also for the oxides, which are often associated by -OH and -OH_2. The CEC for oxides and hydroxides is dependent on pH, where a high pH favors the H^+ ions to dissociate from the functional groups and to replace the vacant places with cations. The oxides and hydroxides are efficient sorbents and sinks for:

- Inorganic ions such as silicate, phosphate, and molybdate.
- Organic anions and molecules such as citrate, fulvic and humic acids.
- Cations such as Al, Cu, Pb, V, Co, Cr, and Ni, some of which are essential for plant growth.

9 Silicon Dioxide

The chemical compound silicon dioxide, also known as silica (from the Latin *silex*), is an oxide of silicon with a chemical formula of SiO_2 and has been known for its hardness since antiquity Silica is most commonly found in nature as sand or quartz, as well as in the cell walls of diatoms. It is a principal component of most types of glass and substances such as concrete. Silica is the most abundant mineral in the earth's crust.

Manufactured Forms

Silica is manufactured in several forms including:

- Glass (a colorless, high-purity form is called *fused silica*)
- Synthetic amorphous silica, silica gel
- Fumed silica (also known as pyrogenic silica, colloidal silica, or under the genericized trademark, Cab-o-Sil)
- Precipitated silica is produced by precipitation from a waterglass solution by acidification.
- Silica aerogel.

Applications

It is used in the production of various products:

- Inexpensive soda-lime glass is the most common and typically found in drinking glasses, bottles, and windows.
- A raw material for many whiteware ceramics such as earthenware, stoneware and porcelain.
- A raw material for the production of Portland cement.
- A food additive, primarily as a flow agent in powdered foods, or to absorb water (see the ingredients list for).
- The natural ("native") oxide coating that grows on · silicon is hugely beneficial in microelectronics. It is a superior electric insulator, with high chemical stability. In electrical applications, it can protect the silicon, store charge, block current, and even act as a controlled pathway to allow small currents to flow through a device. At room temperature, however, it grows extremely slowly, and so to manufacture such oxide layers, the traditional method has been heating of silicon in high-temperature furnaces within an oxygen ambient (thermal oxidation).
- Raw material for aerogel in the Stardust spacecraft.
- Used in the extraction of DNA and RNA due to its ability to bind to the nucleic acids under the presence of chaotropes.
- As hydrophobic silica it is used as a defoamer component.
- As hydrated silica in toothpaste (abrasive to remove plaque.)
- As a high-temperature thermal protection fabric.
- In cosmetics for its light-diffusing properties.
- Liquid silicon dioxide (colloidal silica) is used as a wine and juice fining agent.

- As a glidant in pharmaceutical products silicon dioxide aids powder flow when tablets are formed.
- In the production of tires (Conti Eco Contact 3)
- Thermal enhancement compound used in thermal grouts for the ground source heat pump industry.

Inhalation Health Effects

Inhaling finely divided crystalline silica dust or fumed silica in very small quantities (OSHA allows 0.1 mg/m^3) over time can lead to silicosis, bronchitis or (much more rarely) cancer, as the dust becomes lodged in the lungs and continuously irritates them, reducing lung capacities (silica does not dissolve over time). This effect can be an occupational hazard for people working with sandblasting equipment, products that contain powdered crystalline silica and so on. Children, asthmatics of any age, allergy sufferers and the elderly (all of whom have reduced lung capacity) can be affected in much shorter periods of time.

Other Health Effects

In respects other than inhalation, pure silicon dioxide is inert and harmless.

Pure silicon dioxide produces no fumes and is insoluble *in vivo*. It is indigestible, with zero nutritional value and zero toxicity. When silica is ingested orally, it passes unchanged through the gastrointestinal tract, exiting in the feces, leaving no trace behind. Small pieces of silicon dioxide are equally harmless, as long as they are not large enough to mechanically obstruct the GI tract, or jagged enough to lacerate its lining. (Eating glass and other impure silica is not recommended, however.)

Because some silicas take on water, extended exposure may cause local drying of the skin or other tissue.

According to a French study which followed subjects for 15 years found that higher levels of silica in water appeared to decrease the risk of dementia. The study found that for every 10 miligram-per-day intake of silica in drinking water, the risk of dementia dropped by 11%.

Chemistry

Silicon dioxide is formed when silicon is exposed to oxygen (or air). A very thin layer (approximately 1 nm or 10 Å) of so-called 'native oxide' is formed on the surface when silicon is exposed to air under ambient conditions.

Higher temperatures and alternate environments are used to grow well-controlled layers of silicon dioxide on silicon, for example at temperatures of 600-1200 °C so-called "dry" or "wet" oxidation using O_2 or H_2O respectively The thickness of the layer of silicon replaced by the dioxide is 44% of the thickness of the silicon dioxide layer produced.

Alternative methods used to deposit a layer of SiO_2 include:

- Low temperature oxidation (LTO) of silane

  $SiH_4 + 2O_2 \rightarrow SiO_2 + 2H_2O$ (at 400-450°C)

- Decomposition of *tetraethyl orthosilicate* (TEOS) at 680-730°C

  $Si(OC_2H_5)_4 \rightarrow SiO_2 + H_2O + 2C_2H_4$

- Plasma enhanced chemical vapour deposition using TEOS at approximately 400°C

  $Si(OC_2H_5)_4 + 12O_2 \rightarrow SiO_2 + 10H_2O + 8CO_2$

- Recently, polymerization of *tetraethyl orthosilicate* (TEOS) at less than 100°C using amino acid as catalyst.

Pyrogenic silica (sometimes called *fumed silica* or *silica fume*), which is a very fine particulate form of silicon dioxide, is prepared by burning $SiCl_4$ in an oxygen rich hydrocarbon flame to produce a "smoke" of SiO_2.

$$SiCl_4 + 2H_2 + O_2 \rightarrow SiO_2 + 4HCl$$

Amorphous silica, silica gel, is produced by the acidification of solutions of sodium silicate to produce a gelatinous precipitate that is then washed and then dehydrated to produce colorless microporous silica.

Quartz exhibits a maximum solubility in water at around 340 °C. This property is used to grow single crystals of quartz in a hydrothermal process where natural quartz is dissolved in superheated water in a pressure vessel which is cooler at the top. Crystals of 0.5-1 kg can be grown over a period of 1-2 months. These crystals are a source of very pure quartz for use in electronic applications.

Fluorine reacts with silicon dioxide to form SiF_4 and O_2 whereas the other halogen gases (Cl_2, Br_2, I_2) react much less readily.

Silicon dioxide is attacked by *hydrofluoric acid* (HF) to produce *"hexafluorosilicic acid"*

$$SiO_2 + 6\,HF \rightarrow H_2SiF_6 + 2H_2O$$

HF is used to remove or pattern silicon dioxide in the semiconductor industry.

Silicon dioxide dissolves in hot concentrated alkali or fused hydroxide (e.g):

$$SiO_2 + NaOH \rightarrow Na_2SiO_3 + H_2O$$

Silicon dioxide reacts with basic metal oxides (e.g. sodium oxide, potassium oxide, lead(II) oxide, zinc oxide or mixtures of oxides forming silicates and glasses as the Si-O-Si bonds in silica are broken successively). As an example the reaction of sodium oxide and SiO_2 can produce sodium orthosilicate, sodium silicate and glasses, depending on the proportions of reactants:

$$2Na_2O + SiO_2 \rightarrow Na_4SiO_4$$

$$Na_2O + SiO_2 \rightarrow Na_2SiO_3$$

$$(0.25 - 0.8)Na_2O + SiO_2 \rightarrow \text{glasses}$$

Examples of such glasses have commercial significance e.g. soda lime glass,borosilicate glass, lead glass. In these glasses silica is termed the network former or lattice former.

With silicon at high temperatures gaseous SiO is produced.

$SiO_2 + Si = 2SiO$ (gas)

Structure and Properties

SiO_2 has a number of distinct crystalline forms in addition to amorphous forms. With the exception of stishovite and fibrous silica, all of the crystalline forms involve tetrahedral SiO_4 units linked together by shared vertices in different arrangements. Silicon-oxygen bond lengths vary between the different crystal forms, for example in a-quartz the bond length is 161 pm, whereas in a-tridymite it is in the range 154-171 pm The Si-O-Si angle also varies between a low value of 140° in a-tridymite, up to 180° in ß-tridymite. In a-quartz the Si-O-Si angle is 144°.

Fibrous sulfur has a structure similar to that of SiS_2 with chains of edge-sharing SiO_4 tetrahedra.

Stishovite, the highest pressure form, in contrast has a rutile like structure where silicon is 6 coordinate. The density of stishovite is 4.287 g/cm^3, which compares to a-quartz, the densest of the low pressure forms, which has a density of 2.648 g/cm^3. The difference in density can be ascribed to the increase in coordination as the six shortest Si-O bond lengths in stishovite (four Si-O bond lengths of 176 pm and two others of 181 pm) are greater than the Si-O bond length (161 pm) in a-quartz. The change in the coordination increases the ionicity of the Si-O bond.

Note that the only stable form under normal conditions is a-quartz and this is the form in which crystalline silicon dioxide is usually encountered In nature impurities in crystalline a-quartz can give rise to colours.

Table 9.1: Crystalline forms of SiO_2

Form	Crystal Class	Structural features	Notes
1	2	3	4
α-quartz	rhombohedral (trigonal)	Helical chains making individual single crystals optically active	α-quartz converts to b quartz at 573°C
β-quartz	hexagonal	closely related to a-quartz (with an Si-O-Si angle of 155°) and optically active	β-quartz converts to b-tridymite at 870°C
α-tridymite	orthorhombic	metastable form under normal pressure	
β-tridymite	hexagonal	closely related to a-tridymite	β-tridymite converts to β-cristobalite 1470°C
α-cristobalite	tetragonal	metastable form under normal pressure	
β-cristobalite	cubic	closely related to a-cristobalite	melts at 1705°C
keatite	tetragonal	Si_5O_{10}, Si_4O_{14}, Si_8O_{16} rings	synthesised from amorphous silica and alkali at high pressure

(Contd...)

1	2	3	4
coesite	monoclinic	Si_4O_8 and Si_8O_{10}	high pressure form (higher than keatite)
stishovite	tetragonal	rutile like with 6-fold coordinated Si	high pressure form (higher than coesite) and the densest of the polymorphs
melanophlogile	Cubic	Si_5O_{10}, Si_6O_{17} rings	Mineral always found with hydro-carbons in intestitial spaces – a ceathrasil
fibrous	Orthorhombic	like SiS_2 consisting of edge sharing chairs.	

Molecular Forms of Silicon Dioxide

When molecular silicon monoxide, SiO, is condensed in an argon matrix cooled with helium along with oxygen atoms generated by microwave discharge, molecular SiO2 is produced which has a linear structure. The Si-O bond length is 148.3 pm which compares with the length of 161 pm in a-quartz. The bond energy is estimated at 621.7 kJ/mol.

Dimeric silicon dioxide, $(SiO_2)^2$ has been prepared by reacting O_2 with matrix isolated dimeric silicon monoxide, (Si_2O_2) In dimeric silicon dioxide there are two oxygen atoms bridging between the silicon atoms with an Si-O-Si angle of 94° and bond length of 164.6 pm and the terminal Si-O bond length is 148.2 pm.

COLLOIDAL SILICA

Colloidal silicas are suspensions of fine amorphous, nonporous, and typically spherical silica particles in a liquid phase.

Properties

Usually they are suspended in an aqueous phase that is stabilized electrostatically. Colloidal silicas exhibit particle densities in the range of 2.1 to 2.3 g/cm^3.

Most colloidal silicas are prepared as monodisperse suspensions with particle sizes ranging from approximately 5 to 100 nm in diameter. Polydisperse suspensions can also be synthesized and have roughly the same limits in particle size. Smaller particles are difficult to stabilize while particles much greater than 150 nanometers are subject to sedimentation.

Manufacture

Colloidal silicas are most often prepared in a multi-step process where an alkali-silicate solution is partially neutralized, leading to the formation of silica nuclei. The subunits of colloidal silica particles are typically in the range of 1 to 5 nm. Whether or not these subunits are joined together depends on the conditions of polymerization. Initial acidification of a water-glass (sodium silicate) solution yields $Si(OH)_4$.

If the pH is reduced below 7 or if salt is added, then the units tend to fuse together in chains. These products are often called silica gels. If the pH is kept slightly on the alkaline side of neutral, then the subunits stay separated, and they gradually grow. These products are often called precipitated silica or silica sols. Hydrogen ions from the surface of colloidal silica tend to dissociate in aqueous solution, yielding a high negative charge. Substitution of some of the Si atoms by Al is known increase the negative colloidal charge, especially when it is evaluated at pH below the neutral point. Because of the very small size, the surface area of colloidal silica is very high.

The colloidal suspension is stabilized by pH adjustment and then concentrated, usually by evaporation. The maximum concentration obtainable depends on the on particle size. For example, 50 nm particles can be concentrated to greater than 50 wt% solids while 10 nm particles can only be concentrated to approximately 30 wt% solids before the suspension becomes too unstable.

Applications

- In papermaking colloidal silica is used a a drainage aid. It is capable of increasing the amount of cationic starch that can be retained as a dry-strength agent.
- High temperature binders.
- Investment casting-used in moulds.
- An abrasive - for polishing silicon wafers.
- Carbonless paper.
- Catalysts.
- Abrasion resistant coatings.

- Increasing friction - used to coat waxed floors, textile fibers and railway tracks to promote traction.
- Antisoiling – fills micropores to prevent take up of dirt and other particles into textiles.
- Surfactant – used for flocculating, coagulating, dispersing, stabilising etc.
    - ❑ Liquid silicon dioxide (colloidal silica) is used as a wine and juice fining agent.
- Absorbent.
- Colloidal silica used in concrete densifiers and polished concrete.
- In manufacturing Quantum dots small semi-condutors used in various scientific research settings.

10 Carbonate

In chemistry, a carbonate is a salt or ester of carbonic acid.

Applications

To test for the presence of the carbonate anion in a salt, the addition of dilute mineral acid (e.g. hydrochloric acid) will yield carbon dioxide gas.

Carbonate-containing salts are industrially and mineralogically ubiquitous. The term "carbonate" is also commonly used to refer to one of these salts or carbonate minerals. Most common is calcite, or calcium carbonate, $CaCO_3$, the chief constituent of limestone. The process of removing carbon dioxide from these salts by heating is called calcination.

The term is also used as a verb, to describe the process of raising carbonate and bicarbonate concentrations in soda, see also carbonated water, either by the introduction under pressure of carbon dioxide gas into the bottle, or by dissolving carbonate or bicarbonate salts into the water.

Structure and Bonding

The carbonate ion is a polyatomic anion with the empirical formula CO_3^{2-} and a molecular mass of 60.01 daltons; it consists of one central carbon atom surrounded by three identical oxygen atoms in a trigonal planar arrangement, and has D_{3h} molecular symmetry. The carbonate ion carries a negative two formal charge and is the conjugate base of the hydrogen carbonate ion, HCO_3^-, which is the conjugate base of H_2CO_3, carbonic acid.

The structure and bonding of the carbonate ion cannot be properly represented by its Lewis structure, which depicts CO_3^{2-} with two long single bonds and one short double bond:

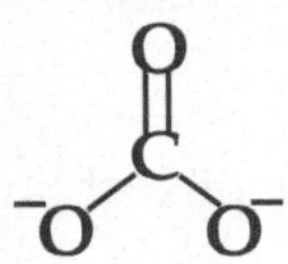

Resonance structures can be used to depict the carbonate ion:

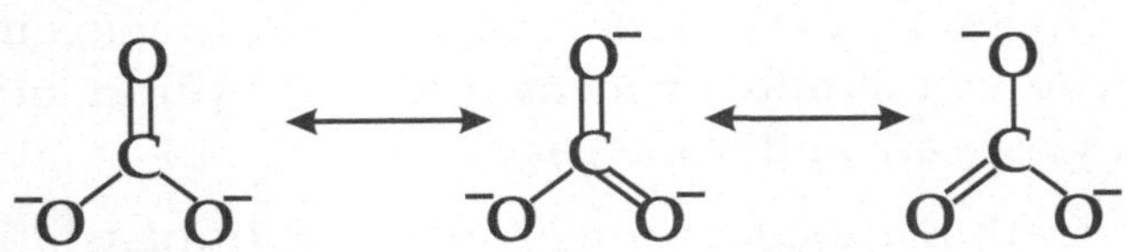

In reality, CO_3^{2-} has three equally long C-O bonds:

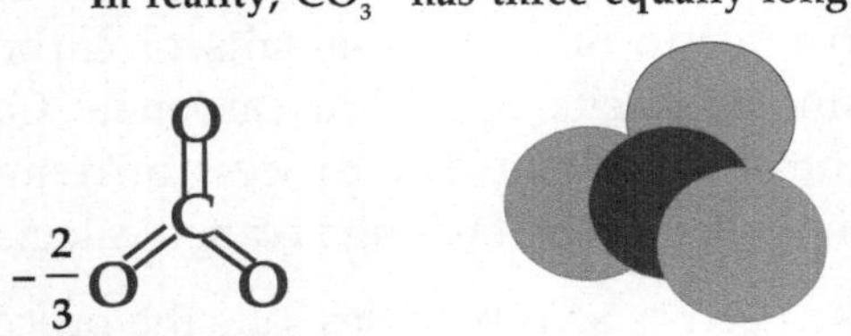

Chemical Properties

A carbonate salt forms when a positively charged ion attaches to the negatively charged oxygen atoms of the ion, forming an ionic compound:

$$2M^{+} + CO_3^{2-} \rightarrow M_2CO_3$$

$$M^{2+} + CO_3^{2-} \rightarrow MCO_3$$

$$2M^{3+} + 3CO_3^{2-} \rightarrow M_2(CO_3)_3$$

Most carbonate salts are insoluble in water at standard temperature and pressure, with solubility constants of less than 1×10^{-8}. Exceptions include sodium, potassium and ammonium carbonates.

In aqueous solution, carbonate, bicarbonate, carbon dioxide, and carbonic acid exist together in a dynamic equilibrium. In strongly basic conditions, the carbonate ion predominates, while in weakly basic conditions, the bicarbonate ion is prevalent. In more acid conditions, aqueous carbon dioxide, $CO_2(aq)$, is the main form, which, with water, H_2O, is in equilibrium with carbonic acid - the equilibrium lies strongly towards carbon dioxide. Thus sodium carbonate is basic, sodium bicarbonate is weakly basic, while carbon dioxide itself is a weak acid.

Carbonated water is formed by dissolving CO_2 in water under pressure. When the partial pressure of CO_2 is reduced, for example when a can of soda is opened, the equilibrium for each of the forms of carbonate (carbonate, bicarbonate, carbon dioxide, and carbonic acid) shifts until the concentration of CO_2 in the solution is equal to the solubility of CO_2 at that temperature and pressure. In living systems an enzyme, carbonic anhydrase, speeds the interconversion of CO_2 and carbonic acid.

Acid-base Chemistry

The carbonate ion (CO_3^{2-}) is a moderately strong base. It is a conjugate base of the weakly acidic bicarbonate (IUPAC name hydrogen carbonate HCO_3^{-}), itself a moderately strong conjugate base of the still weakly acidic carbonic acid. As such in aqueous solution, the carbonate ion seeks to reclaim hydrogen atoms.

Organic Carbonates

In organic chemistry a carbonate can also refer to a functional group within a larger molecule that contains a carbon atom bound to three oxygen atoms, one of which is double bonded. These compounds are also known as organocarbonates or carbonate esters, and have the general formula ROCOOR′, or $RR'CO_3$. Important organocarbonates include dimethyl carbonate, the cyclic compounds ethylene carbonate and propylene carbonate, and the toxic triphosgene.

Biological Significance

It works as a buffer in the blood as follows: when pH is too low, the concentration of hydrogen ions is too high, so you exhale CO_2. This will cause the equation to shift left, essentially decreasing the concentration of H^+ ions, causing a more basic pH.

When pH is too high, the concentration of hydrogen ions in the blood is too low, so the kidneys excrete bicarbonate (HCO_3^-). This causes the equation to shift right, essentially increasing the concentration of hydrogen ions, causing a more acidic pH.

It is generally thought that the presence of carbonates in rock is unchallengable evidence for the presence of liquid water. Recent observations of the Planetary nebula NGC 6302 shows evidence for carbonates in space where aqueous alteration similar to that on Earth is unlikely. Other minerals have been proposed which would fit the observations.

Significant carbonate deposits have not been found on Mars via remote sensing or in situ missions, even though Martian meteorites contain small amounts. Groundwater may have existed at both Gusev and Meridiani Planum.

SIGNIFICANCE IN PEDOLOGY

In horizons with concretions of hard nodules a 'c' is used for iron, aluminium, manganese, or titanium cemented nodules or concretions. A 'g' denotes gleying indicated by low chroma colour (< 2), either total gleying or the presence of gleying in a mottled pattern. In illuvial horizons the accumulation of organic matter with or without sesquioxides (the oxides and hydroxides of iron and aluminium) are denoted by a 'h'. If the sesquioxide component contains enough iron so that colour value and chroma exceeds 3, 'hs' is used. Generally, the accumulation of iron and the cementation (i.e., more than 90% of the horizon is cemented) is denoted by 'sm'. Residual accumulation of sesquioxides after intense weathering is denoted by 'o'.

IRON OXIDES AND HYDROXIDES

Primary minerals which contain iron are for example biotite, pyroxene, amphibole, and olivine. Iron oxides and hydroxides are formed by protonation and release of Fe ions out of primary or secondary minerals and/or oxidation. Their occurence provides useful information about soil formation.

Another important attribute of Fe is that its cationic charge is sensitive to changes in the redox status of the soil. This may impart clues via soil colour about drainage conditions in the soil. Generally, in soils that do not have impeded drainage the majority of the Fe-oxides occur in the Fe^{3+} state, which is typically associated with red, yellow or brown subsoil colours. With a progressive increase in impeded drainage conditions subsoil colours reflect an increasing influence of Fe^{2+} on soil colour. Fe^{2+} typically imparts a bluish gray colour to poorly drained subsoils, often referred to as a gley colour. Along this continuum of drainage conditions variations on patterns may be identified [so-called redoximorphic features (RMF)]. In soils that are intermediate between well-drained and waterlogged, periodic reducing conditions may result in particular zones within the soil exhibiting variegated colour patterns. These patterns consist of intricate combinations of reddish (high chroma RMFs: containing local concentrations of Fe^{3+} oxides), whitish or light grey areas (low chroma RMFs: reflecting zones from which Fe has largely been removed) and sometimes bluish gray areas (gleyed zones: reflecting presence of Fe^{2+}-oxides). The distinction between low chroma zones, which are local zones of Fe eluviation, and gleyed zones is not always simple in the field. Both conditions are indicative of reducing conditions, but represent differences in the extent of Fe^{2+} translocation. Soil colour and in particular patterns of redoximorphic features are used as field indicators of drainage conditions. The criteria are developed to suit local conditions and are ideally based on hydrological and chemical measurements, which are calibrated with observed morphology. This is particularly important because the distribution pattern of Fe-oxide colours may not necessarily reflect contemporary processes operating in the soil. In other words, some of the Fe-oxides still present in the soil may have formed under conditions quite different from those operating at present or they may reflect colours of inherited Fe-oxides or other minerals such as glauconite (greenish colour that could be confused as a gley feature) that have not been significantly altered as a result of pedogenesis.

Where iron oxides are absence, soil colour usually arise from uncoated mineral grains. They may occur evenly dispersed throughout the soil horizons or as concentrations in particular morphological features such as RMFs, nodules, pipestems. Sesquioxides are a term for the oxides of iron and aluminium and sesquioxic coatings (such as ferrans) can be formed by reduction and solution of Fe under anaerobic conditions and their subsequent oxidation and depositon in aerobic zones. If iron oxides (and manganese oxides) become concentrated in a soil horizon they may form cemented layers, called fragipans (denoted by a 'x'), which are hard to very hard and brittle when dry. In contrast, zones of Fe depletion are called neoalbans, which may occur in eluviated soil horizons. The term plinthite is used for B or C horizons (denoted by a 'v'), which are humus poor and iron rich. The material usually has reticulate mottling of reds, yellows, and gray colours and hardens irreversibly to ironstone hardpans or aggregates with repeated wetting and drying.

Iron oxides and hydroxides are very stable under aerobic conditions, but they become more soluble under anaerobic conditions (low redox potentials). They are able to form metal-organic complexes, where the metal cations are bonded by functional groups such as -COOH, =CO, -OH, $-OCH_3$, $-NH_2$, -SH to organic compounds resulting in the formation of a ring structure incorporating the metal ion. These complexes are very stable and called chelates.

Reduction or oxidation can occur outside these boundaries, but only when mediated by an organism and at expense of metabolic energy.

Several different Fe oxides and hydroxides can be distinguished, which differ in their crystal structure and various other properties (e.g. colour, solubility, thermal behavior). The basic unit of iron oxides and hydroxides ia the $Fe(O,OH)_6$ octahedron. The variation between different iron oxides and hydroxides is mainly due to a variation in the arrangement of these octaheda. Following the 6 most widespread Fe oxides and hydroxides are described briefly.

Goethite (alpha-FeOOH)

It is the most frequently occuring Fe-oxide in soil and has a characteristic yellowish brown colour. In large concentrations it may also appear as dark brown or black. Goethite is found under a broad range of climatic and hydrological conditions and is the thermodynamically most stable of all Fe-oxides. A variety of conditions appear to favor its formation in preference to hematite, which include cool temperatures, moist soil conditions and presence of comparatively large amounts of organic matter. The absence of hematite and abundance of goethite in many soils of cool or temperate regions supports these general observations. Under suitable conditions, it has been suggested that goethite can form from any Fe source.

Hematite (alpha-Fe_2O_3)

It has a characteristic bright red colour and primarily occurs in the better drained soils of warm temperate to tropical areas. It is also found in restricted areas of cool temperate areas either as a primary mineral or in more highly weathered soils, where it probably formed under climatic conditions warmer and drier than those operating at present. Its formation appears to be favored by warm dry conditions and small amounts of organic matter (OM). There are large areas of intertropics that have yellow topsoils (OM-rich) over red subsoils, which often contain yellow rims around root channels. This supports the concept of the OM-antihematic effect, i.e., hematite formation is hindered in the presence of comparatively large amounts of OM. Hematite probably forms from ferrihydrite ($5Fe_2O_3*9H_2O$) through aggregation, dehydration, and internal structural rearrangement of the tiny ferrihydrite particles. This has been suggested because ferrihydrite has a hematite-like structure, except that it is highly disordered and is hydrated. As such, ferrihydrite is considered a necessary precursor for hematite formation.

Lepidocrocite (gamma-FeOOH)

It has a characteristic bright orange colour that is evident in soils where it occurs in large concentrations and is not masked by other pigments. It commonly occurs in association with goethite and usually in soils that have restricted drainage. It forms from the oxidation of Fe^{2+} compounds, which are common in wet soils. High partial pressure of CO_2 appears to favor goethite formation in preference to lepidocrocite. This has been demonstrated in laboratory experiments and is supported by field observations that show its virtual absence in calcareous soils, and preferential formation of goethite adjacent to roots and lepidocrocite concentrated further away in wet soils. Lepidocrocite formation also appears to be favored by slow oxidation rates and small concentrations of aluminium in the soil solution. Lepidocrocite is rarely found in very acid soils, which typically contain aluminium in the soil solution.

Maghemite (gamma-Fe_2O_3)

It is common in many soils, notably in the tropics and subtropics and varies in colour from red to brown. This oxide occurs in soils, especially those derived from basic igneous rocks. Maghemites are commonly concentrated towards the top of the soil profile. Several pathways have suggested to account for their formation in soils, which include:

(i) Oxidation of magnetite

(ii) Dehydration of lepidocrocite

(iii) Transformation by heating of other Fe-oxides between 300 and 4250C in the presence of organic compounds.

Magnetite

It is listed in most textbooks as a primary soil mineral that does not form through pedogenic processes. It occurs in soils as singular irregular black grains and is present in most soils. It is fairly resistant to weathering but may alter slowly to ferrihydrite, goethite or eventually hematite (via ferrihydrite). It has been documented the occurrence of magnetic bacteria in soils that contain minute crystals of magnetite. This the first account of in situ 'biogenic' formation of the mineral in soils.

Ferrihydrite ($5Fe_2O_3$*$9H_2O$)

It is a common Fe-oxide found in soils and appears reddish brown in colour if not masked by other pigments. It has a poorly ordered structure resembling that of hematite and was previously referred to as amorphous ferric hydroxide. Its formation is favored by rapid oxidation of Fe in the presence of large concentrations of OM and/or silicate. As such it occurs in limited amounts in tropical soils that do not have these conditions. Phosphate anions also appear to favor its formation in preference to other Fe-oxides. The presence of ferrihydrite generally indicates that conditions are not favorable to crystal growth. However, it is generally considered to be a 'young' Fe-oxide that will on a pedogenic time scale eventually transform to a more stable and more crystalline Fe-oxide (i.e., crystal-inhibiting substances become degraded, translocated etc., allowing for its transformation).

Some other important iron compounds found in soils are:

- *Siderite* ($FeCO_3$): It is found in waterlogged soils, i.e. under reducing environmental conditions. The colour is greenish/blue.
- *Vivianite* ($Fe_3[PO_4]_2{*}H_2O$): It is found in waterlogged soils, i.e. under reducing environmental conditions. The colour is greenish/blue.
- *Pyrit* (FeS): It is found in waterlogged soils, i.e. under reducing environmental conditions. The colour of FeS is black.

ALUMINIUM OXIDES AND HYDROXIDES

The aluminium oxides and hydroxides have a non-distinctive grayish-white colour, which is easily masked in soils except when large concentrations occur. In acid soils, deposits of amorphous aluminium hydroxide form in the interlayers of expanding lattice clays, and occur as surface coatings on clay minerals generally. This amorphous material slowly crystallizes to gibbsite (gamma-$Al(OH)_3$), the principal aluminium hydroxide in soil, which is very stable material. This may become dehydrated to form boehmite (alpha-AlOOH), which is common in bauxite deposits but less common in soils. Gibbsite accumulates in old soils that are in an advanced stage of weathering and in younger soils of the tropics.

Al^{3+} in solution hydrolyzes to produce H^+ as follows:

$Al^{3+} + H_2O \longleftrightarrow Al(OH)^{2+} + H^+$

The hydroxy aluminium ion may also hydrolyze:

$Al(OH)^{2+} + H_2O \longleftrightarrow Al(OH)_2^{+1} + H^+$

Thus, the major source of H+ in moderately and strong acidic soils is aluminium hydrolysis. The Al^{3+} cation has a higher charge than other cations such as K^+, Na^+, Ca^{2+}, or Mg^{2+}, and in soils with pH <5 the cations associated with minerals are replaced by Al^{3+} rather than by H_3O^+. The ability to replace cations is smaller for H_3O^+ in comparison to Al^{3+}.

MANGANESE OXIDES AND HYDROXIDES

The weathering of primary minerals, such as biotite, pyroxene, amphibole, containing Mn^{2+}, produce in an aerobic environment brown/black Mn^{4+}. The reaction can be written as:

$Mn^{2+} + H_2O \rightarrow MnO_2 + 4H^+ + 2e^-$

The pyrolusite (MnO_2) is a very stable manganese oxide. Often manganese is associated with other ions such as Ba, Ca, K, Na, Li, NH_4, Co, Cu and Ni, therefore the manganese oxides and hydroxides have variable forms. For example, birnessite (Na, Ca, K, Mg, Mn^{2+}) $Mn_6^{4+} O_{14}* H_2O$, lithiophorite ($LiAl_2Mn^{2+}Mn_2^{4+}O_9*3H_2O$), or hollandite ($BaMn_8O_{16}$). Manganese oxides show even a greater tendency than iron oxides to occur in concretions. A reason might be the reduction of Mn4+ to Mn^{2+}, which is relatively soluble, more readily soluble than for example Fe^{2+}. Manganese-rich micromorphological zones in peds that are black, often also contain large amounts of iron oxides. However, iron-rich concretions have been shown to be low in manganese oxides. Manganese oxide minerals have a black colour, which is sometimes difficult to distinct from the black colour or organic material.

Other Non-Silicates

Sulphates

Gypsum ($CaSO_4*2H_2O$) which is common in soils (-> fertilizers) and jarosite ($KFe_3(OH)_6(SO_4)_2$ which is rarely found in soils are sulphates. An accumulation of gypsum in a B or C horizon is denoted with a 'y'. If there is a cementation of more than 90% and roots can penetrated only through cracks, the symbol 'ym' is used.

Chlorides

The chloride halite (NaCl) is found in soils formed in marine or lacustrine deposits. The accumulation of sodium is denoted by a 'n', indicating a high level of exchangeable sodium in the horizon.

Carbonates

Primary minerals which contain Ca2+ are calcite ($CaCO_3$), dolomite $(Ca,Mg)CO_3$, plagioclase, pyroxene, and amphibole. They are easily weatherable and Ca^{2+} can be released. The solubility of Ca^{2+} is dependent on the CO_2 partial pressure (pCO_2). The higher the pCO_2 the higher the pH ($pH = -0.67 \lg pCO_2 + 7.23$, where pCO_2 is in kPa) A pH < 7 will be reached if all carbonate is dissolved and begins to leach or migrate. Carbonates have a characteristic to stick mineral particles and organic compounds together to form aggregates. Therefore, carbonates improve soil structure, particularly in A and B horizons. An accumulation of carbonates, usually calcium carbonate, in a horizon is designated with a 'k'. If more than 90% of the horizon is cemented by calcium and roots can penetrate only through cracks, the symbol 'km' is used. If carbonates are leached from the A or even B horizon those layers become acid to very acid. This alters the soil structure, cation exchange capacity, biological activity, etc.

Phosphates

A primary mineral which contains P is apatite (Ca_4(CaF or CaCl)$(PO_4)_3$. Apatite is stable in non acid soils but when the pH drops below 7 apatite is weathered fast. Orthophosphates ($H_2PO_4^-$ and HPO_4^{2-}) that are released by mineralization is rapidly adsorbed by the soil particles. This process is called phosphate fixation because the process is difficult to reverse. Phosphate on surfaces that can be readily desorbed and phorphate in solution is called labile phosphate. In contrast, the P held in insoluble compounds or organic matter is called non-labile phosphate (P occluded in surface oxides and insoluble P compounds). Organic P, which is the major source of P for the soil microorganisms and mesofaunas is rapidly mineralized and/or immobilized by the microorganisms, especially bacteria, which have a relative high P requirement. Bacterial phorphate residues comprise mainly the insoluble Ca, Fe, and Al salts of inositol hexaphosphate, the phytates.

11 Feldspar

Feldspars ($KAISi_3 O_8$ – No Al $Si_3 O_8$ – Ca $Al_2 Si_2 O_8$) are a group of rock-forming tectosilicate minerals which make up as much as 60% of the Earth's crust.

Feldspars crystallize from magma in both intrusive and extrusive igneous rocks, as veins, and are also present in many types of metamorphic rock. Rock formed entirely of plagioclase feldspar is known as anorthosite Feldspars are also found in many types of sedimentary rock.

Etymology

Feldspar is derived from the German *Feld,* field, and *Spat,* a rock that does not contain ore. "Feldspathic" refers to materials that contain feldspar. The alternative spelling, *felspar,* has now largely fallen out of use

Compositions

This group of minerals consists of framework or tectosilicates. Compositions of major elements in common feldspars can be expressed in terms of three endmembers:

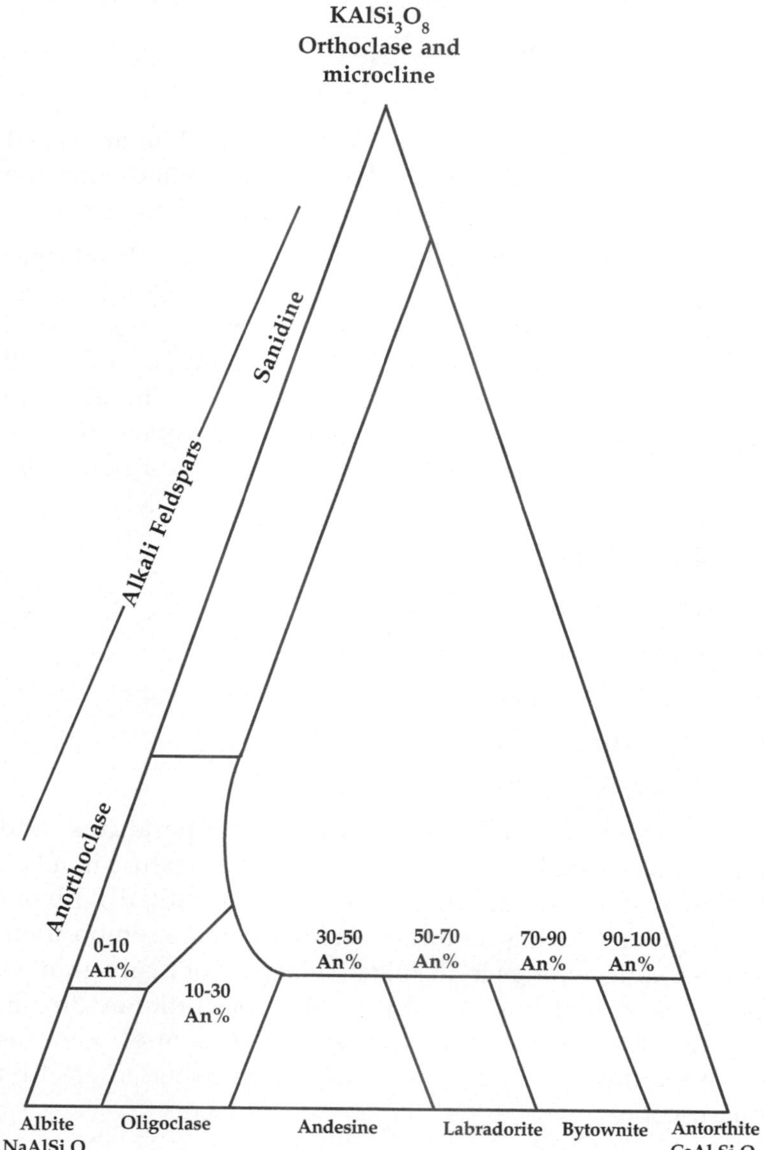

Fig 11.1: **Compositional phase diagram of the different minerals that constitute the feldspar solid solution**

Potassium-Feldspar (K-spar) endmember $KAlSi_3O_8$

Albite endmember $NaAlSi_3O_8$

Anorthite endmember $CaAl_2Si_2O_8$

Solid solutions between K-feldspar and albite are called alkali feldspar. Solid solutions between albite and anorthite are called *plagioclase,* or more properly *plagioclase feldspar.*

Only limited solid solution occurs between K-feldspar and anorthite, and in the two other solid solutions, immiscibility occurs at temperatures common in the crust of the earth. Albite is considered both a plagioclase and alkali feldspar. In addition to albite, barium feldspars are also considered both alkali and plagioclase feldspars. Barium feldspars form as the result of the replacement of potassium feldspar.

Alkali Feldspars

The alkali feldspars are as follows:

- orthoclase (monoclinic), — $KAlSi_3O_8$
- sanidine (monoclinic) — $(K,Na)AlSi_3O_8$
- microcline (triclinic) — $KAlSi_3O_8$
- anorthoclase (triclinic) — $(Na,K)AlSi_3O_8$

Sanidine is stable at the highest temperatures, and microcline at the lowest. Perthite is a typical texture in alkali feldspar, due to exsolution of contrasting alkali feldspar compositions during cooling of an intermediate composition. The perthitic textures in the alkali feldspars of many granites can be seen with the naked eye. Microperthitic textures in crystals are visible using a light microscope, whereas cryptoperthitic textures can only be seen using an electron microscope.

Plagioclase Feldspars

The plagioclase feldspars are triclinic. The plagioclase series follows (with percent anorthite in parentheses):

Fig 11.2: **Potassium feldspar crystals in a granite**

- *albite* (0 to 10) — $NaAlSi_3O_8$
- *oligoclase* (10 to 30) — $(Na,Ca)(Al,Si)AlSi_2O_8$
- *andesine* (30 to 50) — $NaAlSi_3O_8$ — $CaAl_2Si_2O_8$
- *labradorite* (50 to 70) — $(Ca,Na)Al(Al,Si)Si_2O_8$
- *bytownite* (70 to 90) — $(NaSi,CaAl)AlSi_2O_8$
- *anorthite* (90 to 100) — $CaAl_2Si_2O_8$

Intermediate compositions of plagioclase feldspar also may exsolve to two feldspars of contrasting composition during cooling, but diffusion is much slower than in alkali feldspar, and the resulting two-feldspar intergrowths typically are too fine-grained to be visible with optical microscopes. The immiscibility gaps in the plagioclase solid solution are complex compared to the gap in the alkali feldspars. The play of colours visible in some feldspar of labradorite composition is due to very fine-grained exsolution lamellae.

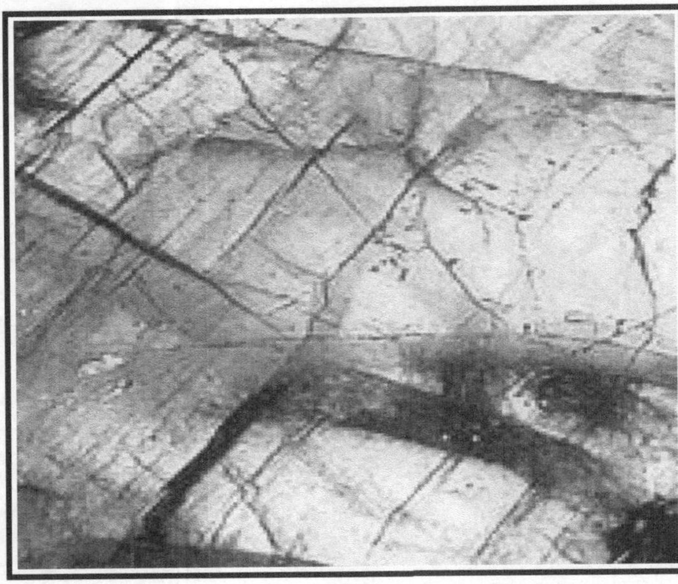

Fig 11.3: **Labradorite**

Barium Feldspars

The barium feldspars are monoclinic and comprise the following:

- celsian — $BaAlSi_3O_8$
- hyalophane — $(K,Na,Ba)(Al,Si)_4O_8$

Feldspars can form clay minerals through chemical weathering.

Uses

- Feldspar is a common raw material in the production of ceramics and geopolymers.
- Feldspars are used for thermoluminescence dating and optical dating in earth sciences and archaeology.
- Feldspar is one of several abrasive ingredients in Bon Ami, a brand of household cleaner in the USA.

In 2005, Italy was the top producer of feldspar with almost one-fifth world share followed by Turkey, China and Thailand, reports the International Monetary Fund.

Why is Feldspar Important?

The feldspars are a family of silicate minerals which occur in igneous rocks. There are many different members to the feldspar group. Obviously, silica and oxygen form the foundation for the group, but calcium, sodium, and potassium are also present. One of these elements is usually dominant, but most of the feldspars contain all 3 in varying amounts. It is the proportions of these 3 elements which help determine which specific feldspar is formed. The feldspars are divided into 2 broad categories: plagioclase, which contains calcium and sodium; and orthoclase, which contains potassium. The plagioclase feldspars represent the "continuous branch" of Bowen's Reaction Series, and form a complete series between anorthite (the pure calcium member), and albite (the sodium-rich variety).

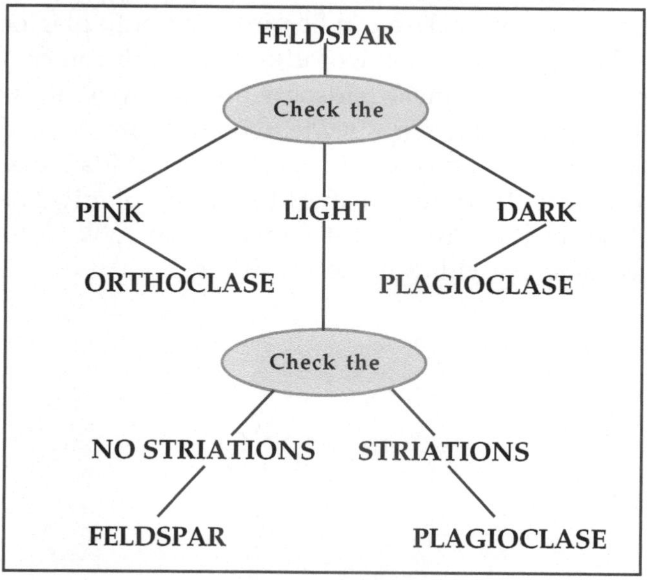

Fig. 11.4

Sixty per cent of the earth's crust is composed of feldspar. This is quite a number, and since feldspar is nearly always present in igneous rocks, most classification schemes depend on the amount and kind of feldspar. In general terms, mafic and intermediate rocks contain plagioclase, with the more calcium present, the more mafic the resulting rock. Orthoclase occurs only in the felsic igneous rocks.

So if we're faced with the need to name an igneous rock, it is important to be able to distinguish between plagioclase and orthoclase. This is obviously harder than it should be, and a final determination is often impossible without chemical and/or optical studies in a lab. We can, however, make an educated guess in the field based on several assumptions, and a few easily identifiable physical features - colour and the presence (or absence) of "striations." Striations are often visible on plagioclase, and resemble very fine (almost microscopic) parallel lines cut into the face of a mineral fragment. All feldspars which have striations are plagioclase, but not all plagioclase has striations. Only orthoclase can be pink, and only plagioclase can be dark gray to black. Unfortunately, both can be light in colour, and since most feldspars are nearly white we're often still faced with a nasty identification problem. If it's white and has striations, I call it plagioclase. If it's white but can't see any striations, it may be either plagioclase or orthoclase. In this case, call it feldspar and look for other clues to come up with a name for the rock.

12 Metamorphism

Metamorphism is the solid-state recrystallization of pre-existing rocks due to changes in physical and chemical conditions, primarily heat, pressure, and the introduction of chemically active fluids. Both mineralogical, chemical and crystallographic changes can occur during this process.

Three types of metamorphism exist: dynamic, contact and regional. Metamorphism produced with increasing pressure and temperature conditions is known as prograde metamorphism. Conversely, decreasing temperatures and pressure characterize retrograde metamorphism.

Limits of Metamorphism

The temperature lower limit of metamorphism is considered to be between 100-150°C, to exclude diagenetic changes, due to compaction, which result in sedimentary rocks. There is no agreement as for a pressure lower limit. Some workers argue that changes in atmospheric pressures are not metamorphic, but some types of metamorphism can occur at extremely low pressures.

The upper boundary of metamorphic conditions is related to the onset of melting processes in the rock. The maximum temperature for metamorphism is typically between 700-900°C, depending on the pressure and on the composition of the rock. Migmatites are rocks formed at this upper limit, which contain pods and veins of material that has started to melt but has not fully segregated from the refractory residue. Since the 1980s, it has been recognized that rarely, rocks are dry enough, and of a refractory enough composition, to record without melting "ultra-high" metamorphic temperatures of 900-1100°C.

Metamorphic Facies

Metamorphic facies are recognizable terranes or zones with an equilibrium assemblage of key minerals that were in equilibrium under specific range of temperature and pressure during a metamorphic event. The facies are named after the metamorphic rock formed under those facies conditions from basalt. Facies relationships were first described by Pentti Eskola in 1921.

Facies

- Low T - low P: Zeolite
- Mod - high T - low P: Prehnite-Pumpellyite
- High-P low T: Blueschist
- Mod P - Mod to high T: Greenschist - Amphibolite - Granulite
- High P - Mod - high T : Eclogite

Metamorphic Grades

In the Barrovian sequence (described by George Barrow in zones of progressive metamorphism in Scotland), metamorphic grades are also classified by mineral assemblage based on the appearance of key minerals in rocks of pelitic (shaly, aluminous) origin:

Low grade ———— Intermediate ——— High grade

Greenschist ——— Amphibolite ———— Granulite

Slate — Phyllite —— Schist ———— Gneiss ———— *Migmatite* (partial melting) >>>melt

Chlorite zone

Biotite zone

Garnet zone

Staurolite zone

Kyanite zone

Sillimanite zone

Contact (Thermal) Metamorphism

Contact metamorphism occurs typically around intrusive igneous rocks as a result of the temperature increase caused by the intrusion of magma into cooler country rock. The area surrounding the intrusion (called *aureoles*) where the contact metamorphism effects are present is called the *metamorphic aureole*. Contact metamorphic rocks are usually known as hornfels. Rocks formed by contact metamorphism may not present signs of strong deformation and are often fine-grained.

Contact metamorphism is greater adjacent to the intrusion and dissipates with distance from the contact. The size of the aureole depends on the heat of the intrusive, its size, and the temperature difference with the wall rocks. Dikes generally have small aureoles with minimal metamorphism whereas large ultramafic intrusions can have significantly thick and well-developed contact metamorphism.

The metamorphic grade of an aureole is measured by the peak metamorphic mineral which forms in the aureole. This is usually related to the metamorphic temperatures of pelitic or alumonisilicate rocks and the minerals they form.

The metamorphic grades of aureoles are andalusite hornfels, sillimanite hornfels, pyroxene hornfels.

Magmatic fluids coming from the intrusive rock may also take part in the metamorphic reactions. Extensive addition of magmatic fluids can significantly modify the chemistry of the affected rocks. In this case the metamorphism grades into metasomatism. If the intruded rock is rich in carbonate the result is a skarn. Fluorine-rich magmatic waters which leave a cooling granite may often form greisens within and adjacent to the contact of the granite. Metasomatic altered aureoles can localize the deposition of metallic ore minerals and thus are of economic interest.

Hydrothermal Metamorphism

Hydrothermal metamorphism is the result of the interaction of a rock with a high-temperature fluid of variable composition. The difference in composition between existing rock and the invading fluid triggers a set of metamorphic and metasomatic reactions. The hydrothermal fluid may be magmatic (originate in an intruding magma), circulating groundwater, or ocean water. Convective circulation of water in the ocean floor basalts produces extensive hydrothermal metamorphism adjacent to spreading centers and other submarine volcanic areas. The patterns of this hydrothermal alteration is used as a guide in the search for deposits of valuable metal ores.

Shock Metamorphism

This kind of metamorphism occurs when either an extraterrestrial object (a meteorite for instance) collides with the Earth's surface or during an extremely violent volcanic eruption. Impact metamorphism is, therefore, characterized by ultrahigh pressure conditions and low temperature. The resulting minerals (such as SiO_2 polymorphs coesite and stishovite) and textures are characteristic of these conditions.

Dynamic (Cataclastic) Metamorphism

Dynamic metamorphism is associated with major fault planes. Metamorphism is localised adjacent to the fault plane and is caused by frictional heat generated by the fault movement. Cataclasis, crushing and grinding of rocks into angular fragments, occurs in dynamic metamorphic zones, giving cataclastic texture.

The textures of dynamic metamorphic zones are dependent on the depth at which they were formed, as the temperature and confining pressure determine the deformation mechanisms which predominate. Within depths less than 5 km, dynamic metamorphism is not often produced because the confining pressure is too low to produce frictional heat. Instead, a zone of breccia or cataclasite is formed, with the rock milled and broken into random fragments. This generally forms a mélange. At depth, the angular breccias transit into a ductile shear texture and into mylonite zones.

Within the depth range of 5-10 km pseudotachylite is formed, as the confining pressure is enough to prevent brecciation and milling and thus energy is focused into discrete fault planes. The frictional heating in this case may melt the rock to form pseudotachylite glass, and adjacent to these zones, result in growth of new mineral assemblages.

Within the depth range of 10-20 km, deformation is governed by ductile deformation conditions and hence frictional heating is dispersed throughout shear zones, resulting in a weaker thermal imprint and distributed deformation. Here, deformation forms mylonite, with dynamothermal metamorphism observed rarely as the growth of porphyroblasts in mylonite zones.

Overthrusting may juxtapose hot lower crustal rocks against cooler mid and upper crust blocks, resulting in conductive heat transfer and localised contact metamorphism of the cooler blocks adjacent to the hotter blocks, and often retrograde metamorphism in the hotter blocks. The metamorphic assemblages in this case are diagnostic of the depth and temperature and the throw of the fault and can also be dated to give an age of the thrusting.

Prograde and Retrograde Metamorphism

Metamorphism is further divided into prograde and retrograde metamorphism. Prograde metamorphism involves the change of mineral assemblages (paragenesis) with increasing temperature and (usually) pressure conditions. These are solid state dehydration reactions, and involve the loss of volatiles such as water or carbon dioxide. Prograde metamorphism results in a rock representing the maximum pressure and temperature experienced. These rocks often return to the surface without undergoing retrograde metamorphism , where the mineral assemblages would become more stable under lower pressures and temperatures.

Retrograde metamorphism involves the reconstitution of a rock under decreasing temperatures (and usually pressures) where revolatisation occurs; allowing the mineral assemblages formed in prograde metamorphism to return to more stable minerals at the lower pressures. This is a relatively uncommon process, because volatiles must be present for retrograde metamorphism to occur. Most metamorphic rocks return to the surface as a representation of the maximum pressures and temperatures they have undergone.

METAMORPHIC ROCK

Metamorphic rock is the result of the transformation of an existing rock type, the *protolith*, in a process called metamorphism, which means "change in form". The protolith is subjected to heat and pressure (temperatures greater than 150 to 200 °C and pressures of 1500 bars causing profound physical and/or chemical change. The protolith may be sedimentary rock, igneous rock or another older metamorphic rock. Metamorphic rocks make up a large part of the Earth's crust and are classified by texture and by chemical and mineral assemblage (metamorphic facies). They may be formed simply by being deep beneath the Earth's surface, subjected to high temperatures and the great pressure of the rock layers above it. They can be formed by tectonic processes such as continental collisions which cause horizontal pressure, friction and distortion. They are also formed when rock is heated up by the intrusion of hot molten rock called magma from the Earth's interior.

The study of metamorphic rocks (now exposed at the Earth's surface following erosion and uplift) provides us with very valuable information about the temperatures and pressures that occur at great depths within the Earth's crust.

Some examples of metamorphic rocks are gneiss, slate, marble, schist, and quartzite.

Metamorphic Minerals

Metamorphic minerals are those that form only at the high temperatures and pressures associated with the process of metamorphism. These minerals, known as index minerals, include sillimanite, kyanite, staurolite, andalusite, and some garnet.

Other minerals, such as olivines, pyroxenes, amphiboles, micas, feldspars, and quartz, may be found in metamorphic rocks, but are not necessarily the result of the process of metamorphism. These minerals formed during the crystallization of igneous rocks. They are stable at high temperatures and pressures and may remain chemically unchanged during the metamorphic process. However, all minerals are stable only within certain limits, and the presence of some minerals in metamorphic rocks indicates the approximate temperatures and pressures at which they were formed.

The change in the particle size of the rock during the process of metamorphism is called *recrystallization*. For instance, the small calcite crystals in the sedimentary rock limestone change into larger crystals in the metamorphic rock marble, or in metamorphosed sandstone, recrystallisation of the original quartz sand grains results in very compact quartzite, in which the often larger quartz crystals are interlocked. Both high temperatures and pressures contribute to recrystallization. High temperatures allow the atoms and ions in solid crystals to migrate, thus reorganizing the crystals, while high pressures cause solution of the crystals within the rock at their point of contact.

Foliation

The layering within metamorphic rocks is called *foliation* (derived from the Latin word *folia*, meaning "leaves"), and it occurs when a rock is being compressed along one axis during recrystallization. This causes the platy or elongated crystals of minerals, such as mica and chlorite, to grow with their long axes perpendicular to the orientation of the force. This results in a banded, or foliated, rock, with the bands showing the colors of the minerals that formed them.

Textures are separated into *foliated* and *non-foliated* categories. Foliated rock is a product of differential stress that deforms the rock in one plane, sometimes creating a plane of cleavage: for example, slate is a foliated metamorphic rock, originating from shale. Non-foliated rock does not have planar patterns of stress.

Rocks that were subjected to uniform pressure from all sides, or those which lack minerals with distinctive growth habits, will not be foliated. Slate is an example of a very fine-grained, foliated metamorphic rock, while phyllite is coarse, schist coarser, and gneiss very coarse-grained. Marble is generally not foliated, which allows its use as a material for sculpture and architecture.

Another important mechanism of metamorphism is that of chemical reactions that occur between minerals without them melting. In the process atoms are exchanged between the minerals, and thus new minerals are formed. Many complex high-temperature reactions may take place, and each mineral assemblage produced provides us with a clue as to the temperatures and pressures at the time of metamorphism.

Metasomatism is the drastic change in the bulk chemical composition of a rock that often occurs during the processes of metamorphism. It is due to the introduction of chemicals from other surrounding rocks. Water may transport these chemicals rapidly over great distances. Because of the role played by water, metamorphic rocks generally contain many elements that were absent from the original rock, and lack some which were originally present. Still, the introduction of new chemicals is not necessary for recrystallization to occur.

Types of Metamorphism

Regional Metamorphism

Regional or Barrovian metamorphism covers large areas of continental crust typically associated with mountain ranges, particularly subduction zones or the roots of previously eroded mountains. Conditions producing widespread regionally metamorphosed rocks occur during an orogenic event. The collision of two continental plates or island arcs with continental plates produce the extreme compressional forces required for the metamorphic changes typical of regional metamorphism. These orogenic mountains are later eroded, exposing the intensely deformed rocks typical of their cores. The conditions within the subducting slab as it plunges toward the mantle in a subduction zone also produce regional metamorphic effects. The techniques of structural geology are used to unravel the collisional history and determine the forces involved. Regional metamorphism can be described and classified into metamorphic facies or metamorphic zones of temperature/pressure conditions throughout the orogenic terrane.

Contact Metamorphism

Contact metamorphism is the name given to the changes that take place when magma is injected into the surrounding solid rock (country rock). The changes that occur are greatest wherever the magma comes into contact with the rock because the temperatures are highest at this boundary and decrease with distance from it. Around the igneous rock that forms from the cooling magma is a metamorphosed zone called a *contact metamorphism aureole*. Aureoles may show all degrees of metamorphism from the contact area to unmetamorphosed (unchanged) country rock some distance away. The formation of important ore minerals may occur by the process of metasomatism at or near the contact zone.

When a rock is contact altered by an igneous intrusion it very frequently becomes more indurated, and more coarsely crystalline. Many altered rocks of this type were formerly called hornstones, and the term *hornfels* is often used by geologists to signify those fine grained, compact, non-foliated products of contact metamorphism. A shale may become a dark argillaceous hornfels, full of tiny plates of brownish biotite; a marl or impure limestone may change to a grey, yellow or greenish lime-silicate-hornfels or siliceous marble, tough and splintery, with abundant augite, garnet, wollastonite and other minerals in which calcite is an important component. A diabase or andesite may become a diabase hornfels or andesite hornfels with development of new hornblende and biotite and a partial recrystallization of the original feldspar. Chert or flint may become a finely crystalline quartz rock; sandstones lose their clastic structure and are converted into a mosaic of small close-fitting grains of quartz in a metamorphic rock called quartzite.

If the rock was originally banded or foliated (as, for example, a laminated sandstone or a foliated calc-schist) this character may not be obliterated, and a banded hornfels is the product; fossils even may have their shapes preserved, though entirely recrystallized, and in many contact-altered lavas the vesicles are still visible, though their contents have usually entered into new combinations to form minerals which were not originally present. The minute structures, however, disappear, often completely, if the thermal alteration is very profound; thus small grains of quartz in a shale are lost or blend with the surrounding particles of clay, and the fine ground-mass of lavas is entirely reconstructed.

By recrystallization in this manner peculiar rocks of very distinct types are often produced. Thus shales may pass into cordierite rocks, or may show large crystals of andalusite (and chiastolite), staurolite, garnet, kyanite and sillimanite, all derived from the aluminous content of the original shale. A considerable amount of mica (both muscovite and biotite) is often simultaneously formed, and the resulting product has a close resemblance to many kinds of schist. Limestones, if pure, are often turned into coarsely crystalline marbles; but if there was an admixture of clay or sand in the original rock such minerals as garnet, epidote, idocrase, wollastonite, will be present. Sandstones when greatly heated may change into coarse quartzites composed of large clear grains of quartz. These more intense stages of alteration are not so commonly seen in igneous rocks, because their minerals, being formed at high temperatures, are not so easily transformed or recrystallized.

In a few cases rocks are fused and in the dark glassy product minute crystals of spinel, sillimanite and cordierite may separate out. Shales are occasionally thus altered by basalt dikes, and feldspathic sandstones may be completely vitrified. Similar changes may be induced in shales by the burning of coal seams or even by an ordinary furnace.

There is also a tendency for metasomatism between the igneous magma and sedimentary country rock, whereby the chemicals in each are exchanged or introduced into the other. Granites may absorb fragments of shale or pieces of basalt. In that case hybrid rocks called skarn arise which have not the characters of normal igneous or sedimentary rocks. Sometimes an invading granite magma permeates the rocks around, filling their joints and planes of bedding, etc., with threads of quartz and feldspar. This is very exceptional but instances of it are known and it may take place on a large scale.

Regional Metamorphism

Regional metamorphism is the name given to changes in great masses of rock over a wide area. Rocks can be metamorphosed simply by being at great depths below the Earth's surface, subjected to high temperatures and the great pressure caused by the immense weight of the rock layers above. Much of the lower continental crust is metamorphic, except for recent igneous intrusions. Horizontal tectonic movements such as the collision of continents create orogenic belts, and cause high temperatures, pressures and deformation in the rocks along these belts. If the metamorphosed rocks are later uplifted and exposed by erosion, they may occur in long belts or other large areas at the surface. The process of metamorphism may have destroyed the original features that could have revealed the rock's previous history. Recrystallization of the rock will destroy the textures and fossils present in sedimentary rocks. Metasomatism will change the original composition.

Regional metamorphism tends to make the rock more indurated and at the same time to give it a foliated, shistose or gneissic texture, consisting of a planar arrangement of the minerals, so that platy or prismatic minerals like mica and hornblende have their longest axes arranged parallel to one another. For that reason many of these rocks split readily in one direction along mica-bearing zones (schists). In gneisses, minerals also tend to be segregated into bands; thus there are seams of quartz and of mica in a mica schist, very thin, but consisting essentially of one mineral. Along the mineral layers composed of soft or fissile minerals the rocks will split most readily, and the freshly split specimens will appear to be faced or coated with this mineral; for example, a piece of mica schist looked at facewise might be supposed to consist entirely of shining scales of mica. On the edge of the specimens, however, the white folia of granular quartz will be visible. In gneisses these alternating folia are sometimes thicker and less regular than in schists, but most importantly less micaceous; they may be lenticular, dying out rapidly. Gneisses also, as a rule, contain more feldspar than schists do, and they are tougher and less fissile. Contortion or crumbling of the foliation is by no means uncommon, and then the splitting faces are undulose or puckered. Schistosity and gneissic banding (the two main types of foliation) are formed by directed pressure at elevated temperature, and to interstitial movement, or internal flow arranging the mineral particles while they are crystallizing in that directed pressure field.

Rocks which were originally sedimentary and rocks which were undoubtedly igneous are converted into schists and gneisses, and if originally of similar composition they may be very difficult to distinguish from one another if the metamorphism has been great. A quartz-porphyry, for example, and a fine feldspathic sandstone, may both the converted into a grey or pink mica-schist

Metamorphic Rock Textures

The five basic metamorphic textures with typical rock types are:

- *Slaty:* slate and phyllite; the foliation is called 'slaty cleavage'
- *Schistose:* schist; the foliation is called 'schistosity'
- *Gneissose:* gneiss; the foliation is called 'gneissosity'
- *Granoblastic:* granulite, some marbles and quartzite
- *Hornfelsic:* hornfels and skarn.

13 Rocks and Minerals

Actinolite

Actinolite has the formulae $Ca_2(Mg,Fe)_5 Si_8O_2 2(OH)_2$ and a relative hardness of 6. It commonly occurs in crystalline schists, often being the chief constituent of green-coloured schists and greenstones. It is used somewhat as an asbestos material.

Adamantine

Adamantine refers to a brilliant luster like that of a diamond. Results from a mineral having a high index of refraction.

Adamite

Adamite has the formulae $Zn_2(AsO_4)(OH)$ and a relative hardness of 4. It has brilliant fluorescence, is of scientific interest and is much desired by collectors.

Aegirite

Aegirite has the formulae $NaFe(Si_2O_6)$ and a relative hardness of 7. It is a rock-forming mineral found mainly in rocks rich in soda and poor in silica. Named afterAegir, Icelandic god of the sea.

Agaric Mineral

Agaric Mineral is one of the purest native carbonates of lime. It is chiefly found in the clefts of rocks and at the bottom of some lakes.

Agate

Agate is a semi-precious stone comprised mainly of silica. It is a compact variety of chalcedony. It has a pattern of banding which occurs due to chemical admixtures of hematite, limonite and other minerals.

Alabaster

Alabaster is naturally occurring hydrated calcium sulphate.

Albite

Albite has the formulae $NaAlSi_3O_8$ and a relative hardness of 7. It is widely distributed, rock-forming mineral. Present in pegmatite dikes and may be found in crystals.

Alkaline Rock

Alkaline rock refers to any rock which contains more than average amounts of potassium bearing and sodium bearing minerals.

Allanite

Allanite has the formulae $(Ce,Ca,Y)_2(Al,Fe)_3(SiO_4)_3(OH)$ and a relative hardness of 6. It occurs as a minor constituent of many igneous rocks and is frequently associated with epidote. Found in some magnetic bodies.

Allemontite

Allemontite is a natural alloy of arsenic and antimony. It has a relative hardness of 4.

Almandine

Almandine is a variety of precious garnet, reddish or violet in colour.

Alteration

Alteration refers to physical or chemical change in a rock or mineral after its original formation. Can result in new minerals or in textural changes in the rock.

Alumina

Alumina is aluminium oxide and occurs as ruby, sapphire, and bauxite.

Aluminum

Aluminium is a light-weight silvery metal element with the symbol Al.

Alunite

Alunite has the formulae $KAl_3(SO_4)2(OH)6$ and a relative hardness of 4. It is usually formed by sulphuric acid solutions acting on rocks rich in alkalic feldspar. Used in the production of alum.

Amazonite

Also called Amazonstone, amazonite is a green microcline with the formula $KaISI_3O_8$.

Amber

Amber is fossilised resin.

Amblygonite

Amblygonite has the formulae (Li,Na)Al(PO4)(F,OH). It has a relative hardness of 6. It occurs in lithium and phosphate rich granite pegmatites. Associated with spodumene, apatite, tourmaline, lepidolite.

Amethyst

Amethyst is a precious stone. It is quartz coloured with manganese.

Amorphous

Amorphous refers to 'Without form'. The term is applied to rocks and minerals that lack definite crystal structure.

Amphibole

The name of a mineral group or "family." All the members have as their basic structure the formula Si_4O_{11}. Hornblende is one of the most common rock forming minerals of the group.

Amygdule

Amygdule refers to a mineral containing cavity in an igneous rock formed by escaping gas.

Anaclime

Anaclime has the formulae $NaAlSi_2O_6$ù$_2H_2O$ and a relative hardness of 6.

It is found in the cavities of intrusive and volcanic igneous rocks; often as clear shiny crystals and is associated with calcite and zeolites.

Anatase

Anatase has the formulae TiO_2 and a relative hardness of 6. It is a form of rutile found in granite, gneiss, mica schist, maetamorphic limestone, and dolomite. May be present as an accessory mineral in the rocks or in a quartz vein traversing it. It is used as a coating for welding rods and as a derivative of titanium.

Andalusite

Andalusite has the formulae Al_2SiO_5. It has a relative hardness of 8. It is formed by the metamorphism of aluminous shales and slate. Used in the manufacture of spark plugs and other porcelains. May serve as a gem stone if clear and transparent. Name from Andalusia, a province of Spain.

Andesine

Andesine has the formulae $NaAlSi_3O_8$. It has a relative hardness of 6.

Anglesite

Anglesite is a common, minor ore of lead formed by the oxidation of galena.It has the formulae $PbSO_4$ and a relative hardness of 3. It is found in the upper, oxidzed portions of lead veins and is named after the Island of Anglesey.

Anhydrite

Anhydrite has the formulae $CaSO_4$. It has a relative hardness of 4. It has rectangular cleavage fragments. Occurs in a manner similar to gypsum and often found with it but not as common. Found in the cap rock of salt domes and in limestone rocks. From the Greek meaning 'without water'.

Annivite

Annivite is a variety of terahedrite containing bismuth and usually iron and zinc.

Anorthite

Anorthite has the formulae $(Na,Ca)AlSi_3O_8$. It has a relative hardness of 7. It is widely distributed and abundant rock-forming feldspar.

Anthophyllite

Anthophyllite has the formulae $(Mg,Fe)_7Si_8O_{22}(OH)_2$. It has a relative hardness of 6. It occurs in crystalline schists rich in magnesium. Of interest to collectors.

Antimonite

Antimonite is an antimony ore. It was used by the Greeks as a cosmetic for darkening the eyelids.

Antimony

Antimony has the formulae Sb. It has a relative hardness of 4. It is a native semimetal and difficult to distinguish from arsenic. Rather brittle and a poorer conductor of heat and electriciy than native metals.

Apatite

Apatite has the formulae $Ca_5(F,Cl, OH)(PO_4)_3$. It has a relative hardness of 5. It is widely distributed as an accesory constituent in all classes of rock. Found in crystals and masses. Crystallized apatite has been extensivley for fertilizer. Transparent varieties of fine colour are sometimes used for gems - but it's too soft for extensive use.

Aphanitic Rock

Aphanitic rock refers to a rock in which the crystalline constituents are too small to be distinguished without magnification.

Apophyllite

Apophyllite has the formulae $KCa_4Si_8O_2O(OH)ù_8H_2O$. It has a relative hardness of 5. It is a hydrothermal mineral filling cavities in basalt and tufaceous rocks. Associated with stilbite, scolecite, calcite, prehnite, analcime.

Aragonite

Aragonite has the formulae $CaCO_3$. It has a relative hardness of 4. It forms easily in a sedimentary environment but only stable in metamorphic rocks formed at high temperatures.

Arfvedsonite

Arfvedsonite has the formulae Na2-3 $(Fe,Mg,Al)_5$ Si_8O_{22} $(OH,F)_2$. It has a relative hardness of 6. It is typical of alkaline plutonic rocks that are rich in iron.

Argentite

Argentite has the formulae Ag_2S. It has a relative hardness of 3. It lacks distinct cleavage. An important primary silver ore. The name is from Latin meaning 'silver'. Occurs disseminated in galena and in the cementation zone of lead and zinc deposits. Associated with cerussite, chlorargyrite, native silver.

Argillaceous

Argillaceous refers to composed mostly of or containg clay; such as shale.

Arsenates

Arsenates refers to minerals in which arsenate (AsO4) is an important part.

Arsenic

Arsenic is a metal element with the symbol As and a relative hardness of 4. It has a garlicky odor; otherwise it is difficult to distinguish from antimony. It is a relatively rare mineral found in veins in crystalline rocks associated with silver, cobalt, or nickel ores.

Arsenopyrite

Arsenopyrite has the formulae FeAsS. It has a relative hardness of 6. It is the most common mineral containing arsenic. It occurs with tin and tungsten ores in high temperature deposits as a deposition from hot waters. Frequently associated with gold. Also found with copper and silver ores.

Asbestos

Asbestos is a fibrous mineral with a high melting and combustion point. Cloth was made of it by the ancient Egyptians.

Atacamite

Atacamite has the formulae $Cu_2Cl(OH)_3$. It has a relative hardness of 4. It is a comparatively rare copper mineral. Occurs in arid regions in the upper oxidized zone of copper deposits.

Augite

Augite has the formulae $(Ca,Na)(Mg,Fe,Al,Ti)(Si,Al)2O_6$. It has a relative hardness of 6. It is an important rock forming mineral. Chiefly found in dark coloured igneous rocks, especially those whose magmas were rich in iron, calcium and magnesium. Seldom found in rocks that contain much quartz. A common memeber of the pyroxene group. Told from amphibole by cleavage.

Autunite

Autunite has the formulae $Ca(UO_2)_2(PO_4)_2$ù$10\text{-}12H_2O$. It has a relative hardness of 3. It has strong fluorescence. A secondary mineral found in the zone of oxidation and weathering of uranite or other uranium minerals. Used as an ore of uranium.

Axinite

Axinite has the formulae Ca_2 (Fe,Mn) $Al_2(BO_3)(Si_4O_{12})(OH)$. It has a relative hardness of 7. It occurs in cavities in granite and in the contact zones surrounding granite intrusions.

Azurite

Azurite has the formulae $Cu_3(CO_3)_2(OH)_2$. It has a relative hardness of 4. It is a minor ore of copper with an intense azure-blue colour. Alters to malachite. Associated with limonite, calcite, chalcocite, chrysocolla and other secondary copper minerals. Reacts vigorously with hydrochloric acid.

Barite

Barite has the formulae $BaSO_4$. It has a relative hardness of 3. It is major ore of barium. Has a high specific gravity for a light coloured mineral. Common gangue mineral in hydrothermal veins or as a repacement mineral in veins of limestone and dolomite. Associated with lead, silver and antimony sulphides.

Barium

Barium is a metal element of the alkaline earth group with the symbol Ba.

Barytes

Barytes was the first mineral to be found to be luminescent when heated, and led to the discovery of the luminescence of minerals. It has a relative hardness of 3.5.

Basalt

Basalt is the name given to lava. It is high in ferrous and magnesian silicates.

Basanite

Basanite is a smooth, black siliceous mineral.

Basic Rock

Basic rock refers to an igneous rock with a low percentage of silica and a high percentage of pyroxene, hornblende, and labradorite.

Bauxite

Bauxite is a sedimentary rock group of various aluminium oxides, a principal ore of aluminium, found in France and Jamaica. Bauxite was named after the place where it was first found, Les Baux in France. It has a relative hardness of 1 to 3.

Bedding

Bedding refers to the arrangement of sedimentary rocks in about parallel layers or strata which correspond to the original sediments.

Beryl

Beryl (beryllium aluminium meta-silicate) has the formulae $Be_3Al_2Si_{16}O_{18}$ and a relative hardness of 8. It has a characteristic six-sided outline and is used as a gem stone of various colours, its green variety being emerald. Beryl is also the major source of the rare element beryllium, a light metal similar to aluminum. Beryl is quite common and occurs usually in granite rocks, mica schists and with tin ores.

Beryllium

A rare, lightweight metal element found in the earth's crust with the symbol Be.

Beryllium aluminium meta-silicate

See "Beryl" above.

Biotite

Biotite has the formulae $K(Mg,Fe)_2(Al,Fe)Si_3O_{10}(OH,F)_2$ and a relative hardness of 3. It is a widely distributed rock forming mineral and occurs in igneous and metamorphic rocks and is a common member of the mica group.

Bismuth

Bismuth is a metal element with the symbol Bi and a relative hardness of 3. Bismuth is a somewhat rare mineral occurring in connection with ores of silver, cobalt, nickel, lead, and tin. Bismuth forms low melting alloys and is used in fuses, medicine and cosmetics.

Bismuthinite

Bismuthinite has the formulae Bi_2S_3 and a relative hardness of 2. It is a rare ore of bismuth. Occurs in veins that show definite relations to igneous rocks.

Bituminous Rocks

Bituminous rocks refers to rocks that contain tar, petroleum, or asphalt.

Blende

Blende is a zinc ore. It generally contains more than half its weight in zinc, a quarter sulphur and often a small amount of iron.

Bloodstone

Bloodstone is a dark green variety of chalcedony sprinkled with red jesper hence the name.

Bole

Bole is an earthy mineral occuring in amorphous masses, and composed chiefly of silica with alumina, iron and occasionally magnesia.

Boracite

Boracite has the formulae $Mg_3B_7O_{13}Cl$ and a relative hardness of 7. It occurs associated with beds of halite, anhydrite, and gypsum. It is formed by the evaporation of bodies of salt water.

Borates

Borates refers to a group of minerals in which the borate radical (BO_3) is an important constituent.

Borax

Borax (sodium tetraborate) is the sodium salt of pyroboric acid. It has the formulae $Na_2B_4O_{710}H_2O$ and a relative hardness of 3. It forms large transparent six-sided prisms which have an alkaline reaction, effloresce in air, and when heated swell-up and melt to a transparent glass. Borax is used in the manufacture of enamel-ware, glass, as an antiseptic and is a food preservative.

Bornite

Bornite has the formulae Cu_5FeS_4 and a relative hardness of 3. It is an ore of copper with a colourful tarnish, widely occurring, it is found in basic rocks and metamorphic deposits.

Boron

Boron is an element with the symbol B found in borax.

Botryoidal

Botryoidal refers to resembling a bunch of grapes. A mineral of this type appears to have a surface covered with spherical bulges.

Boulangerite

Boulangerite is a lead ore. It contains 55 percent lead. It has a relative hardness of 2.5.

Bournonite

Bournonite has the formulae $PbCuSbS_3$. It has a relative hardness of 3. It is an ore of lead, copper, and antimony which often exhibits twinned crystals. Occurs in in veins formed at moderate temperatures.

Brass

An alloy of copper and tin discovered during the Bronze Age (6,000-1,000 BC). Harder and more durable than copper by itself and more easily workable than most other known metals and alloys of the period. It was used in making tools, weapons and decorative articles.

Brimstone

Brimstone is another name for sulphur.

Bronze

Bronze is an alloy of copper, zinc and tin.

Brookite

Brookite is a source of titanium but deposits are usually too small to be of commercial use. It has the formulae TiO_2 and a relative hardness of 6.

Brown Spar

Brown Spar is the name given to some crystalline varieties of dolomite tinged with peroxide of Iron

Brucite

Brucite is a decomposition product of silicates, especially serpentine. It has the formulae $Mg(OH)_2$ and a relative hardness of 3. It is found in Texas, where it is white with a grey, green or blue tinge and is used in sugar-refining.

Cadmium

Cadmium is a metal element with the symbol Cd.

Calamine

Calamine is a zinc ore.

Calcareous

Calcareous refers to containing calcium carbonate or calcite.

Calcic

Calcic refers to containing calcium.

Calcite

Calcite has the formulae $CaCO_3$. It has a relative hardness of 3. It effervesces vigorously with hydrochloric acid (HCl). Clear specimens exhibit double refraction. Occurs as widespread sedimentary rock masses such as limestone. Crystalline metamorphosed limestones are called marbles. Fine grained deposits form chalk. Variety: onyx.

Calcium

Calcium is a naturally occurring element with the symbol Ca.

Caliche

Caliche is naturally occurring sodium nitrate found in Chile.

Cannel Coal

Cannel Coal is a dull black coal which breaks with a conchoidal fracture and does not soil the fingers when handled. It is easily cut, and will take a high polish. It contains a large proportion of volatile constituents making it suitable for gas manufacture, and it burns with a bright white flame.

Carbon

Carbon is a non-metallic element with the symbol C. Contained in all life forms.

Carbonaceous

Carbonaceous refers to composed chiefly of organic carbon. (i.e. carbon derived from plant and animal remains.)

Carbonates

Carbonates refers to minerals, such as calcite, where the carbonate radical (CO_3) is an important constituent.

Carnallite

Carnallite is a source of potassium coumpounds and magnesium. Has a bitter, salty taste and has the formulae $KMgCl_{3.6}H_2O$ and a relative hardness of 3.

Carnelian

Carnelian is a clear red chalcedony, a semi-precious gemstone, consisting of quartz with iron impurities which give it a translucent red colour. Carnelian is found mainly in Brazil, Japan and India.

Carnotite

Carnotite has the formulae $<K_2(UO_2)2(VO_4)_{2.3}H_2O$. It has a relative hardness of 1. It is an ore of uranium and vandium. Found in sand, sandstones and around petrified trees. Probably formed as a deposition from meteoric waters. Strongly radioactive.

Cassiterite

Cassiterite has the formulae SnO_2 and a relative hardness of 7. It is the principal ore of tin. Widely distributed in small amounts but commerically available in only a few localities. Frequently associated with wolframite. Also found as rolled pebbles in placer deposits ('stream tin'). Usually found in veins associated with quartz, in or near granitic rocks.

Celestine

Celestine is a natural sulphate of strontium.

Celestite

Celestite has the formulae $SrSO_4$. It has a relative hardness of 4. It often found disseminated through limestone or sandstone, or lining cavities in such rocks. Associated with calcite, dolomite, gypsum, sulphur, fluorite. Also found as a gangue mineral in lead veins. Used to prepare nitrate of strontium for fireworks and tracer bullets and in the refining of beet sugar.

Cerium

Cerium is a rare metal element with the symbol Ce.

Cerussite

Cerussite has the formulae $PbCO_3$. It has a relative hardness of 4. It is an important and widely distributed supergene lead ore formed by the action of carbonated waters on galena in the upper zone of lead veins. Often found associated with galena and sphalerite.

Chabazite

Chabazite has the formulae $Ca(Al_2Si_4)O_{12.6}H_2O$. It has a relative hardness of 5. It is a mineral of secondary origin found lining cavities in volcanic and intrusive igneous rocks.

Chalcanthite

Chalcanthite has the formulae $CuSO_{4.5}H_2O$. It has a relative hardness of 3. It dissolves quickly in water and produces a blue solution. A minor ore of copper and found only in arid regions. Occurs near the surface in copper veins. Often deposited on iron from the water in copper mines. Used in calico printing, insecticides and for industrial purposes.

Chalcedony

Chalcedony is a variant of quartz comprised of silica. Chalcedony was named after Chalkedon, near Istanbul, Turkey. It was traditionally used for decorative objects and amultes. It has a relative hardness of 7.

Chalk

Chalk is a pure soft limestone, opaque white, and usually formed by the accumulation of the shells of foraminifera together with those of larger marine organisms.

Chalcocite

Chalcocite has the formulae Cu_2S. It has a relative hardness of 3. It is one of the most important copper ore minerals. It occurs primarily in enriched zones of sulphide deposits.

Chalcopyrite

Chalcopyrite has the formulae $CuFeS_2$. It has a relative hardness of 4. It is the most widely occuring copper mineral and one of the most important ore sources of that metal. Occurs as an original constituent of igneous rocks, in pegmatic dikes, and in contact with metamorphic deposits. May carry gold or silver and become an ore of those metals.

Chalcosine

Chalcosine is an important copper ore. It was discovered in the 16th century.

Chert

A mineral in which silica was precipitated in sea water or in which silica replaced existing limestone resulting in a very hard (hardness 7) and durable material. Flint is a form of chert. Used from the stone age until the mid-1800's for making tools, weapons and for starting fire when struck with steel.

Chili Saltpetre

See "Cubic Nitr" below.

Chlorite

Chlorite has the formulae $(Mg, Fe)_6(AlSi_3)O_{10}(OH)_8$. It has a relative hardness of 3. It is a mineral group whose members usually exhibit a charactersitic green colour. The formula above is for 'green mica'. Distinguished from muscovite and green phlogopite by a lack of elasticity.

Chondrodite

Chondrodite has the formulae $(Mg,Fe)_3(SiO_4)(OH,F)_2$. It has a relative hardness of 7. It is a common metamorphic mineral in dolomitic marbles.

Chromite

Chromite has the formulae $FeCr_2O_4$. It has a relative hardness of 6. It is slightly magnetic. The only ore of chromium. A common constituent of peridotite rocks and the serpentines derived from them. Also associated with corundum. One of the first minerals to separate from a cooling magma. Chromium is widely used in metal plating and in stainless steel.

Chromium

Chromium is a metal element with the symbol Cr.

Chrysoberyl

Chrysoberyl has the formulae $BeAl_2O_4$. It has a relative hardness of 9. It occurs in granite rocks, pegmatites, and in mica schists. Frequently in river sand and gravels. Serves as a gem stone: alexandrite and "cats eye" which can be of great value.

Chrysocolla

Chrysocolla has the formulae $(Cu,Al)_2H_2Si_2O_5(OH)_4$ nH_2O. It has a relative hardness of 4. It is a minor ore of copper. A mineral of secondary origin, occuring in the oxidized zones of copper veins. Associated with malachite, azurite, cuprite, native copper. Named from two Greek words meaning 'gold' and 'glue'.

Chrysostom

Chrysostom is a gem stone of beryllium aluminate.

Cinnabar

Cinnabar is a red crystalline form of mercuric sulphide. It has the formulae HgS and a relative hardness of 3. It often has a bright red colour. The only important source of mercury and it's found in few localities. Occurs as vein fillings near recent volcanic rocks and hot springs. Used in scientific equipment, in drugs, and with tin in silvering mirrors. Many other uses.

Clastic Rock

Clastic rock refers to a sedimentary rock comprised of fragments of preexisting rocks that have been transported and deposited.

Clay

Clay refers to a soft sediment or deposit that is plastic when wet and comprised of very fine-grained materials, mainly hydrous aluminum silicates.

Cleavage

Cleavage refers to a mineral is said to possess cleavage if when it breaks it yields definite plane surfaces. Cleavage can be perfect as in micas or, in some minerals, completely lacking. Cleavage is always parallel to crystal faces.

Cleveite

Cleveite is a uranium-containing mineral, of interest owing to the fact that when heated with dilute sulphuric acid it liberates considerable quantities of occluded helium. It was this property which first led to the discovery of the element helium.

Clinozoisite

Clinozoisite has the formulae $Ca_2Al_3Si_3O_{12}(OH)$. It has a relative hardness of 7.

Coal

Coal is a solid combustible material of vegetable origin occurring in a fossilised state.

Coal-tar

Coal-tar is a thick black viscous liquid produced by the destructive distillation of coal. Coal-tar yields benzene, creosote, paraffin, aniline and dyes.

Cobalt

Cobalt is a metal element with the symbol Co. It was discovered among the ore veins in Cornwall in early times and called mundic by the miners. It was identified as a metal in 1733 by Brandt.

Cobalt Glance

See "cobaltite" below.

Cobaltite

Cobaltite (cobalt glance) is an ore of cobalt consisting of a cobalt arsenosulphide of the formulae CoAsS and has a relative hardness of 6. It is usually found in high temperature deposits, disseminated in metamorphic rocks, or in vein deposits with other cobalt and nickel minerals.

Coelestine

Coelestine is a mineral used in pyrotechnics and a s a source of strontium. Occassionaly it is cut and used as a gem stone. It has a relative hardness of 3.5.

Colemanite

Colemanite is a major source of borax and has a relative hardness of 5. It occurs in high temperature hydrothermal veins or disseminated in metamorphic rocks associated with other cobalt and nickel sulphides/arsenides.

Columbite

Columbite has the formulae (Fe,Mn)(Nb,Ta)2O6 and a relative hardness of 6. It shows a bluish iridescent fracture surface. The main ore of niobium and tantalum; used in metallurgy to create heat-resistant alloys and in the rust proofing of stainless steel.

Concretion

Concretion refers to an accumulation of mineral matter when mineral particles become cemented together into an orderly, rounded form.

Contact Metamorphism

Contact metamorphism refers to metamorphism resulting from the intrusion of magma which takes place at or near the contact point with the molten rock.

Copper

Copper has the formulae Cu and a relative hardness of 3. It is one of the essential metals of modern civilization. Native copper is found in copper veins but copper sulphides are the principal source ores of the metal.

Cordierite

Cordierite has the formulae $Mg_2Al_4Si_5O_{18}$. It has a relative hardness of 8. It is found as an accessory mineral in granite, gneiss, schists, and in contact metamorphic zones. Transparent specimens of good colour have been used as a gem.

Corundum

Corundum has the formulae Al_2O_3. It has a relative hardness of 9. It is common as an accessory mineral in metamorphic rocks and as an original constituent of certain igneous rocks. Color differences give rise to several varieties of gem, notably, ruby and saphire. The deep red ruby is one of the most valuable gems, second only to emerald and diamond.

Covellite

Covellite has the formulae CuS. It has a relative hardness of 2. It has an indigo-blue colour. Not an abundant material but is found in most copper deposits, usually as a coating in the zone of sulphide enrichment.

Cristobalite

Cristobalite has the formulae SiO_2. It has a relative hardness of 7. It is present in many siliceous volcanic rocks as a lining in cavities. Upon heating to 1470 C it becomes nearly transparent. On cooling it assumes its initial white translucent appearance.

Crocoite

Crocoite has the formulae $PbCrO_4$. It has a relative hardness of 3. It is a rare mineral found in the oxidized zones of lead deposits where lead veins have traversed rocks containing chromite. Not abundant enough to be of commercial value although it does contain chromium. Name is Greek and means ' saffron', an allusion to the colour.

Cross-stone

see "Harmotome".

Cryolite

Cryolite has the formulae Na_3AlF. It has a relative hardness of 3. It often found enclosing brown siderite and grey galena. Always occurs in pegmatites where it's probably a precipitate from flouride rich solutions. Used in the manufacture of sodium salts, certain kinds of glass and porcelain, and as a flux for cleaning metal surfaces.

Crystal

Crystal refers to a solid mineral having a regular geometric shape and bounded by smooth flat surfaces (called crystal faces).

Crystal Symmetry

Crystal symmetry refers to the repetitive pattern of crystal faces caused by the orderly internal arrangements of atoms within a mineral.

Cubic Nitre

Cubic Nitre (sodium nitrate, Chili Saltpetre) is a mineral found mainly in the Tarapaca district of Chile.

Cuprite

Cuprite has the formulae Cu_2O. It has a relative hardness of 4. It is an important ore of copper. Found in the upper oxidized portions of copper veins. Commonly found in crystal forms.

Danburite

Danburite has the formulae $CaB_2Si_2O_8$. It has a relative hardness of 7.

Datolite

Datolite has the formulae $CaBSiO_4(OH)$. It has a relative hardness of 6. It is a mineral of secondary origin found usually in cavities in basalt lavas and similar rocks.

Delvauxite

Delvauxite is a mineral source of phosphate. It has a relative hardness of 2.5.

Destinezite

Destinezite is a mineral. It is a source of phosphate. It has a relative hardness of 3.

Detrital Sediment

Detrital sediment refers to deposited rock and mineral fragments.

Diamond

Diamond has the formulae C. It has a relative hardness of 10. It is the hardest naturally occuring mineral and the most important of the gemstones. Occurs in pale shades of several colours, deep shades are rare. Gem quality diamond is transparent, of attractive colour, and without internal fractures or inclusions of other materials.

Diaspore

Diaspore has the formulae $AlO(OH)$. It has a relative hardness of 7. It is a constituent of bauxites and a major source of aluminum.

Dike

Dike refers to a wall-like body of igneous rock that cuts across layers of surrounding rocks.

Diopside

Diopside has the formulae $CaMgSi_2O_6$ and a relative hardness of 6.

It is usually found as a contact metamorphic mineral in crystalline limestones. Transparent varieties have been cut and used as gemstones.

Dioptase

Dioptase has the formulae $CuSiO_2(OH)_2$ and a relative hardness of 5.

It is found in the oxidation zone of copper deposits, in cavities of massive copper minerals.

Diorites

Diorites are a group of rocks composed essentially of a soda-lime felspar and hornblende, embracing a wide range of types from acid to basic. Diorites were formed by cooling far below the surface and occur in the Scottish Highlands, the Channel Islands, California and other places.

Dolomite

Dolomite has the formulae $CaMg(CO_3)_2$ and a relative hardness of 4. It's powder reacts vigorously with hydrochloric acid (HCl). A potential ore of magnesium. Occurs in widely extended rock masses as dolomitic limestone. Often intimately mixed with calcite. Formed from ordinary limestone by the replacement of calcium by magnesium.

Dumortierite

Dumortierite has the formulae Al7(BO3)(SiO4)3O3 and a relative hardness of 7. It often has a bright colour and fibrous habit. Occurs in metamorphic rocks rich in aluminum. Also in pegmatites and contact metamorphic rocks.

Dysodile

Dysodile is a yellow or green foliated mineral found in limestone.

Emerald

Emerald is a green precious stone variety of the mineral beryl.

Emery

Emery is an impure fine-grained aluminium oxide with the formulae Al_2O_3 employed widely as an abrasive. Typically emery consists of 60 percent corundum and 40 percent iron oxide in the form of magnetite.

Enargite

Enargite has the formulae Cu_3AsS_4. It has a relative hardness of 4. Vein and replacement deposits associated with pyrite, shpalerite, bornite, galena, chalcocite. Used as an ore of copper.

Enstatite

Enstatite has the formulae $Mg_2Si_2O_6$. It has a relative hardness of 6. It is a common and widespread mineral. Found in mafic plutonic and vocanic rocks and often in both metallic and stony meteorites.

Epidote

Epidote has the formulae Ca_2(Al,Fe) 3Si3O12(OH). It has a relative hardness of 7. It is a common mineral. Found in regional and metamorphic rocks of mafic composition and as a product of alteration of other minerals. Sometimes used as a gemstone.

Epsomite

Epsomite has the formulae $MgSO_4$ $7H_2O$. It has a relative hardness of 3. It has a bitter, salty taste; dissolves easily in water. Occurs in delicate fibrous and capillary aggregates. Often called 'epsom salt'. Found as an efflorescent deposit on the walls of caves and sometimes in lake deposits.

Erbium

Erbium is a metal element with the symbol Er.

Erythrite

Erythrite has the formulae $Co_3(AsO_4)_2$ $8H_2O$. It has a relative hardness of 3. It often shows a vivid colour and occurs in thin crusts. An important indicator of cobalt mineral deposits.

Euclase

Euclase has the formulae $BeAlSiO_4(OH)$. It has a relative hardness of 8. It is a variety of berly.

Eudialyte

Eudialyte is a rare and complex silicate mineral easily soluble in acids.

Eudialyte crystals are trigonal and appear in prismatic or table-top shapes, often pink in colour they may also be yellowish-brown. Eudialyte was confirmed as a distinct mineral in 1819, but its chemical formula is still in doubt.

Eurite

Eurite is an acid, igneous rock of the granite group, which consists of quartz and feldspar with muscovite or garnet as accessory minerals, and is mostly found as dykes and veins traversing granite and crystalline schists.

Extravasation

In geology, extravasation describes the issue of lava and other volcanic products from the earth.

Fahlband

A fahlband is a stratum in crystalline rock, containing metallic sulphides.

Fahlerz

Fahlerz or gray copper ore, is of a steel-gray or iron-black colour mineral that occurs crystallized in the form of the tetrahedron and also massive and disseminated. Its fracture is uneven or imperfectly conchoidal. It consists of from 30 to 40 per cent of copper with iron and sulphur; but it also contains in very variable proportions zinc, lead, antimony, and silver.

Fahlunite

Fahlunite or Falunite, named after Fahlun (Falun) in Sweden, is a hydrated silica of alumina, resulting from the alteration of iolite. It is a mineral of a greenish colour, occurring in six-sided prisms.

Faluns

The faluns is a series of strata, of the Middle Tertiary period, of France, abounding in shells, and used by Sir Charles Lyell as the type of his Miocene subdivision.

Famp

In geology, a famp is an indurated wavy calcareous shale found among limestone rocks.

Farcilite

Farcilite was an old mineralogical term for a conglomerate.

Fassaite

Fassaite, named after the Fassa valley in the Tyrol, is a variety of pyroxene.

Faujasite

Faujasite, named after the French geologist Faujas de Saint-Fond is a silicate occurring in square octahedrons together with black augite in the mandelstein of the Kaiserstuhl in Baden.

Fault

In geology, a fault is a dislocation caused by a slipping of rock masses along a plane of facture. The term also describes the dislocated structure resulting from such slipping. The surface along which the dislocated masses have moved is called the fault plane. When this plane is vertical, the fault is a vertical fault; when its inclination is such that the present relative position of the two masses could have been produced by the sliding down, along the fault plane, of the mass on its upper side, the fault is a normal, or gravity, fault. When the fault plane is so inclined that the mass on its upper side has moved up relatively, the fault is then called a reverse (or reversed), thrust, or over thrust, fault. If no vertical displacement has resulted, the fault is then called a horizontal fault. The linear extent of the dislocation measured on the fault plane and in the direction of movement is the displacement; the vertical displacement is the throw; the horizontal displacement is the heave. The direction of the line of intersection of thefault plane with a horizontal plane is the trend of the fault. A fault is a strike fault when its trend coincides approximately with the strike of associated strata (i.e., the line of intersection of the plane of the strata with a horizontal plane) ; it is a dip fault when its trend is at right angles to the strike; an oblique fault when its trend is oblique to the strike. Oblique faults and dip faults are sometimes called cross faults. A series of closely associated parallel faults are sometimes called stepfaults and sometimes distributive faults.

Favosites

Favosites (favosite) is a genus of fossil zoophyta resembling a honeycomb in appearance and common in the Silurian strata of Norway and Wales.

Fayalite

Fayalite, named after the island of Fayal, is a black, greenish, or brownish mineral of the chrysolite group. It is a silicate of iron.

Feather Alum

Feather alum or halotrichite, is a hydrous sulphate of alumina, resulting from volcanic action, and from the decomposition of iron pyrites.

Feather Ore

Feather ore is a sulphide of antimony and lead, sometimes found in capillary forms and like a cobweb, but also massive. It is a variety of Jamesonite.

Feitsui

Feitsui is the Chinese name for a highly prized variety of pale green jade.

Feldspar

Feldspar or felspar is a name given to a group of minerals, closely related in crystalline form, and all silicates of alumina with either potash, soda, lime, or, in one case, baryta. They occur in crystals and crystalline masses, vitreous in lustre, and breaking rather easily in two directions at right angles to each other, or nearly so. The colours are usually white or nearly white, flesh- red, bluish, or greenish. The group includes the monoclinic (orthoclastic) species orthoclase or common potash feldspar, and the rare hyalophane or baryta feldspar; also the triclinic species (called in general plagioclase) microcline, like orthoclase a potash feldspar;

anorthite or lime feldspar; albite or soda feldspar; also intermediate between the last two species, labradorite, andesine, oligoclase, containing both lime and soda in varying amounts. The feldspars are essential constituents of nearly all crystalline rocks, as granite, gneiss, mica, slate, most kinds of basalt and trachyte, etc. The decomposition of feldspar has yielded a large part ofthe clay of the soil, also the mineral kaolin, an essential material in the making of fine pottery. Common feldspar is itself largely used for the same purpose.

Felsite

Felsite or felstone is a hard, compact, volcanic, fine grained rock, flint-like in fracture, consisting essentially of orthoclase feldspar with occasional grains of quartz occurring in the older geological formations.

Fengite

Fengite is a kind of marble or alabaster, formerly sometimes used for windows on account of its transparency.

Gabbronite

Gabbronite or gabronite is, a mineral, a variety of scapolite, occurring' in masses, whose structure is more or less foliated, or sometimes compact. Its colours are grey, bluish or greenish grey, and sometimes red.

Gabbros

Gabbros are basic rocks solidified under great pressure at considerable depths in the earth's crust. They are coarse-grained and consist of plagioclase feldspar with augite. Sometimes the term is used to include similar rocks composed of the same feldspar with other minerals, and according to the minerals contained they are known as norites, troctolites, eucrites, etc.

Gadolinite

Gadolinite, named after the Russian chemist Gadolin, or ytterbite is a naturally occurring complex silicate containing beryllium, iron and many of the rare earth metals, of which the latter is an important source. The principal rare earths that occur in gadolinite are yttrium and erbium, together with smaller amounts of cerium and lanthanum.It is usually found in dull, amorphous masses disseminated through granite; is black, or very dark green in colour, with a resinous lustre. Gahnite (zinc spinel) is a dark green mineral of the spinel group consisting of zinc aluminium oxide. It is named after J G Gahn the Swedish chemist.

Galena

Galena (lead sulphide) or galenite is virtually the only source of lead and an important ore of silver. It is found both in masses and crystallized in cubes, but sometimes in truncated octahedra and has the formulae PbS and a relative hardness of 3. It is a very common metallic mineral, its colour is bluish-grey, like lead, but brighter; its lustre metallic; texture foliated; fragments cubical; soft, but brittle. When found in veins that show a connection to igneous rocks, it is frequently found with silver minerals. Galena is also found in limestones either as veins or as a replacement deposit.

Galena effervesces with nitric and hydrochloric acids. For the most part it contains about 86.6 per cent of lead and 13.4 of sulphur, generally some silver, and also antimony, zinc, iron, and bismuth. Where the proportion of silver is high it is known as argentiferous galena, and worked with a view to the extraction of this metal. Galena occurs principally in the older or primary rocks, being found in England mainly in the Mountain Limestone (base of the Carboniferous formation). In the United States it is very abundant, the deposit of galena in which the mines of Illinois are situated being extensive and important.

Gangue

A gangue is a mineral substance surrounding a metallic ore in a vein.

Gannister

Gannister is a very siliceous close-grained variety of clay with a poor alkali content found especially under coal seams in the Coal Measures of northern England.

Garnet

Garnet is a widely distributed group with several minerals. They are found in both metamorphic and igneous rocks. The chief use of red transparent garnets are as an inexpensive gem stone, however, much is used as an abrasive material. They have the formulae $A_3B_2(SiO_4)_3$ and a relative hardness of 8. The commonest colour of garnet is red and the lustre is vitreous. The dodecahedron and trapezohedron are the common forms. There are also white, green, yellow, brown, and black varieties. The garnet is a silicate, the bases being alumina lime (grossularite, essonite, or cinnamon stone), or alumina magnesia (pyrope), or alumina iron (almandine), or alumina manganese (spessarite), or iron lime (common garnet, melanite, allochroite), or chromium lime (ouvarovite, the colour emerald green). The garnet was, in part, the carbuncle of the ancients. Garnet is a very common mineral in gneiss and mica slate.

Garnierite

Garnierite is an ore of nickel. It's probably formed as an alteration of nickel bearing peridotites. It has a nice apple green colour and the formulae $(Ni,Mg)_6Si_4O_{10}(OH)_8$. It has a relative hardness of 4. It is named after the French geologist Jules Garnier.

Gault

The Gault is a series of beds of clay and marl in the South of England, between the upper and lower greensand of the Cretaceous period. They vary in colour from a light grey to a dark blue, and are developed chiefly in the neighbourhood of Folkestone (hence oftenc called Folkestone marl).

Gaylussite

Gaylussite, named after the French chemist Gay-Lussac, is a yellowish white, translucent mineral, consisting of the carbonates of lime and soda, with water.

Geanticlinal

Geanticlinal is an upward bend or flexure of a considerable portion of the earth's crust, resulting in the formation of a class of mountain elevations called anticlinoria.

Gems

Gems, or precious stones, are sometimes found crystallized in regular shapes and with a natural polish, more commonly they are of irregular shapes and with a rough coat. The term gem often denotes more particularly a stone that is cut, polished, or engraved, and it also includes pearls and various artificial productions. The first and most valuable class of gems includes diamonds, emeralds, rubies, sapphires, and a few others; the second class includes the amethyst, topaz, garnet, etc; while agate, lapis-lazuli, cornelian, etc, though much used for ornament, can scarcely be called gems.

The fabrication of artificial gems became an important industrial art during the 19th century. The base of one class of imitations is a peculiar kind of glass of considerable hardness, brilliancy and refractive power called paste or strass, which is distinguished from ordinary glass by the presence of 50 per cent of oxide of lead among its constituents. When the strass is obtained very pure it is melted and mixed with substances having a metallic base, generally oxides, which communicate to the mass the most varied colours.

Another class, called semi-stones or doublets, are made by affixing thin slices of real gem to an under part of strass by means of an invisible cement. In some cases an imitation is made by setting uncoloured strass or quartz in jewelry with some coloured 'foil' at the back of it.

At the end of the 19th century attempts were made with a fair measure of success to manufacture true gems by artificial processes. The French chemists Becquerel, Ebelman, Gaudin, Despretz, and others did much in this direction.

In 1858 Deville and Caron communicated to the Academy of Sciences, Paris, a process for the production of a number of gems of the corundum class, such as rubies, sapphires, etc. The process essentially consisted in exposing the fluoride of aluminium, together with a little charcoal and boracic acid, in a plumbago crucible protected from the action of the air, to a white heat for about an hour. Many experiments with a view to producing diamonds artificially have also been made. From hydrocarbons, subjected to a very intense heat and enormous pressure, minute crystals, differing from natural diamonds in no respect save brilliancy, have been produced. Artificial diamonds are now widely used in industry.

In art and archaeology the term gem is usually applied to a precious stone cut or engraved in ornamental designs, or with inscriptions. Stones on which the design is raised above the general surface are called cameos; those having the design sunk below the surface are called intaglios. Early specimens of cut gems are seen in the scarabaei or beetle-shaped signets worn in rings by the ancient Egyptians. Among the Greeks, Etruscans, and Romans gem-sculpture held a high place, reaching its highest point under Augustus. Modern gem-engraving dates from the beginning of the 15th century, the chief seats of the art being Italy and Germany. Rome becoming the headquarters of the seal-engraving art.

The traditional tools of the engraver consist of a lathe, and a series of little rods with heads of different shapes, all of which can be adjusted to the lathe. The axis of the lathe is pierced at the centre with an orifice, into which the tools for cutting the stone are firmly fixed by means of a screw. The engraver wets the extremity of the mounted rod with diamond dust made into a paste with oil (traditionally olive oil), and as the wheel is in motion he applies the stone, firmly cemented to a piece of reed, to the revolving tool. The diamond dust enables the tool to cut into the stone with ease. As the design is frequently very elaborate and of the greatest delicacy, the tools are necessarily multiform. The stones used for cameo-cutting often exhibit layers of different colours, so that the raised design has a tint distinct from the ground. Intaglios are very often executed in transparent stones, and the subjects treated in this manner are more limited in number. They are chiefly such as seals, devices, coats of arms, etc.

Genesee Epoch

The Genesee epoch is a closing subdivision of the Hamilton period in the American Devonian system. It is so called because the formations of this period crop out in Genesee, New York.

Geochronology

Geochronology is the branch of geology concerned with ordering and dating of events in the earth's history, including the origin of the earth itself.

Geochronite

Geochronite is a lead-grey or greyish blue mineral with a metallic lustre, consisting of sulphur, antimony, and lead, with a small proportion of arsenic.

Geode

A geode is a round hollow nodule containing earthy matters, sometimes quartz, sometimes agate. Geodes are found in most volcanic rocks and are formed by water depositing materials in the hollows of these rocks.

Geodynamics

Geodynamics is the branch of geology concerned with the forces and processes, especially on a large- scale, of the earth's interior, particularly as regards their effects on the crust or lithosphere.

Geology

Geology is the scientific study of the origin, history, structure, and composition of the earth.

Geomorphology

Geomorphology or geomorphogeny is the branch of geology that is concerned with the structure, origin, and development of the topographical features of the earth's crust.

Geosynclinal

Geosynclinal is the downward bend or subsidence of the earth's crust, which allows of the gradual accumulation of sediment, and hence forms the first step in the making of a mountain range.

Geosyncline

A geosyncline is a broad elongated depression in the earth's crust containing a great thickness of sediment.

Germanite

Germanite is a mineral consisting of a complex copper arsenic sulphide containing germanium, gallium, iron, zinc, and lead: an ore of germanium and gallium.

Geyserite

Geyserite is a loose hydrated form of silica, a variety of opal, deposited in concretionary cauliflower-like masses, around some hot springs and geysers.

Giallo Antico

Giallo Antico is the Italian name of a kind of fine yellow marble, used in ancient Roman architecture and obtained from Numidia.

Gibbsite

Gibbsite is a primarily secondary mineral named after George Gibbs, an American mineralogist. Gibbsite is a hydrate of alumina and gives off a clay-like smell when rubbed or breathed on. It is an important ore of aluminium. Typically it occurs massive, but rarely crystals are found in flattened, tubular, pseudo-hexagonal shapes. It is mostly white, reddish white or colourless with shades of grey, blue and green and a pearly or vitreous lustre.

Gieseckite

Gieseckite, named after Karl Giesecke is a mineral occurring in greenish grey six-sided prisms and having a greasy lustre.

Girasole

Girasole (girasol) is a variety of opal which is usually milk white, bluish white, or sky blue in colour but in a bright light reflects a reddish colour. The brightest are found in Brazil and Siberia. The name is sometimes bestowed on the Asteria sapphire. One variety is known as the fire opal.

Gismondine

Gismondine, named after it's discoverer, Gismondi, is a native hydrated silicate of alumina, lime, and potash, first noticed near Rome.

Glacial Drift

Glacial drift describes earth and rocks which have been transported by moving ice, land ice, or icebergs.

Glacial Epoch

The glacial epoch was a period during which the climate of the modern temperate regions was polar, and ice covered large portions of the northern hemisphere to the mountain tops.

Glaciated Rocks

Glaciated rocks are those rocks whose surfaces have been smoothed, furrowed, or striated, by the action of ice.

Glacier Tables

Glacier tables are large stones found on glaciers supported on pedestals of ice. The stones attain this peculiar position by the melting away of the ice around them, and the depression of its general surface by the action of the sun and rain. The block, like an umbrella, protects the ice below it from both; and accordingly its elevation measures the level of the glacier at a former period. By and by the stone table becomes too heavy for the column of ice on which it rests, or its equilibrium becomes unstable, whereupon it topples over, and falling on the surface of the glacier defends a new space of ice, and begins to mount afresh.

Glance

Glance is a name given to some sulphides, mostly dark-coloured, which have a brilliant metallic lustre, such as the sulphide of copper which is called copper glance.

Glauberite

Glauberite, named after the 17th century German chemist Glauber, is a bitter salty tasting mineral that dissolves slowly in water. It is a sedimentary mineral formed by the evaporation of saline water. It slowly alters to gypsum when exposed to air. It has the formulae Na2Ca(SO4)2.

Glaucodot

Glaucodot is a metallic mineral having a greyish tin-white colour, and containing cobalt and iron, with sulphur and arsenic.

Glaucophane

Glaucophane is a dark-blue sodium rich rock forming mineral which, like other amphiboles, is poor in silica. It is of interest to petrologists in helping to define the metamorphic conditions which formed the surrounding rock. It has the formulae $Na_2(Mg,Fe)_3Al_2Si_8O_{22}(OH)_2$ and a relative hardness of 7. It often has a blue colour.

Globulite

A globulite is a rudimentary form of crystallite, spherical in shape.

Gmelinite

Gmelinite, named after the German chemist Gmelin, is a rhombohedral zeolitic mineral, related in form and composition to chabazite.

Gneiss

Gneiss refers to a crystalline rock, consisting, like granite, of quartz, feldspar, and mica, but having these materials, especially the mica, arranged in planes, so that it breaks rather easily into coarse slabs or flags. Hornblende sometimes takes the place of the mica, and it is then called hornblendic or syenitic gneiss. Similar varieties of related rocks are also called gneiss.

Goethite

Goethite is an ore of iron and a common mineral. Usually formed as a weathering product of iron bearing minerals. The term 'limonite' is commonly used for earthy yellow and brown goethite. It has the formulae FeO (OH) and a relative hardness of 6.

Gold

Gold is a bright yellow coloured, rare, precious metal element with the formulae Au and a relative hardness of 3 occurring in nature in widely distributed small amounts. The main source of gold is in gold-quartz veins where gold was deposited from mineral bearing solutions. When the veins weather, the gold is separated from the quartz and mechanically settles on the stream floor as a placer deposit. The fineness of gold is measured in carats. 24 carat gold is pure, 22 carat gold is 22 parts gold to two parts other metals and so on. Gold is seldom used for any purpose in a state of perfect purity on account of its softness, but is combined with some other metal to render it harder. Standard gold, or the alloy formerly used for the gold coinage of Britain, consists of twenty-two parts of gold and two of copper (being thus 22 carats fine). Articles of jewelry are made of every degree of fineness up to 22 carats, i.e. 22 parts of gold to 2 of alloy. Gold is one of the most ductile and malleable of all the metals It is one of the heaviest of the metals, and not being liable to be injured by exposure to the air, is well suited for making coins and jewelry. Its ductility and malleability are very remarkable. It may be beaten into leaves so exceedingly thin that one grain in weight will cover 56 square inches. It is also extremely ductile; a single grain may be drawn into a wire 500 feet long, and an ounce of gold covering a silver wire is capable of being extended upwards of 1300 miles. It may also be melted and remelted with scarcely any diminution of its quantity. It is soluble in nitro-muriatic acid or aqua regia, and in a solution of chlorine. Its specific gravity is 19.3, or it is about nineteen times heavier than water.

The alloy of gold and silver is found already formed in nature. It is of a paler yellow than pure gold, while the copper alloy has a colour bordering upon reddish yellow. Palladium, rhodium, and tellurium are also met with as alloys of gold.

Gold has been found in larger or smaller quantities in nearly all parts of the world. It is commonly found in reefs or veins among quartz, and in alluvial deposits. Among the latter may be ranked the deposits in river beds, from which the gold is obtained by dredging. Dredging is even employed where the water has to be brought by artificial means. When gold is in rock, quarrying, crushing, washing, and treatment with mercury are employed. The rock is crushed by machinery, and the crushed material is treated with mercury, which dissolves the gold, forming a liquid amalgam, after which the mercury is volatilized, and the gold left behind. Two other processes are also in use, viz. the chlorination and the cyanide. In the former the gold is transformed into soluble gold chloride, and the metal is obtained from this solution by means of sulphuretted hydrogen. The cyanide process is especially useful for ores containing tellurium, or for ores in a fine state of division, and con^sists in dissolving the metal in potassium cyanide solutions, from which it can be obtained by various means. The gold obtained by these methods always contains silver, from which it is separated by the process known as 'parting'.

In alluvial (or placer) deposits it is extracted by washing, in the form of dust, grains, laminae, or nuggets. After the gravel has been turned over and the nuggets have been taken out, the remainder is washed to recover the finer particles of gold. In washing in the pan - 'panning out' - a quantity of the 'dirt', free from stones, is put into a shallow dish with a slight depression in the middle. It is then mixed with water, and the dish held with one side lower than the other, while by a gentle motion the sand and other lighter bodies are washed over the edge of the pan, and the heavy matters containing the gold remain at the bottom. The 'dolly' or 'tossing tub' is a circular tub for washing rather fine stuff that has been sifted. The 'cradle' consists of a short box or trough two meters long, mounted on a kind of rockers, and slightly inclined to allow the mud to run off. A box, with a bottom of iron plate perforated with holes, is placed over the higher end of the trough. The 'pay dirt' (i.e. gravel or sand containing a sufficient amount of gold to be profitably worked) is thrown into this box, and water is run or poured upon it. The finer portion is thus carried through the holes, and directed by an inclined plate into the trough. The cradle is rocked from side to side, the light matters are carried away by the water, and the particles of gold and other heavy matters lodge behind the 'rimes', or transverse bars of wood, with which the bottom of the trough is fitted, and are afterwards collected.

In quartz mining and the case is similar with the hard, solid 'banket' formation of South Africa that contains tlie gold - the ore to be crushed is first passed through a 'stone-breaker' or 'ore crusher', and is further crushed by the 'stamps' or other grinding mill. The ordinary stamp-battery consists, in its lower part, of a cast-iron 'mortar-box', fitted on one or both sides with a fine screen. At the bottom of this box is a row of iron blocks called 'dies', upon which the stamps, or heavy cylindrical cast-iron blocks, are made to rise and fall by means of cams, being thus kept pounding away at the ore in the mortar-box. A stream of water is admitted, and carries the crushed material through the screens. Mercury is fed into the mortar-boxes in small quantities, and much of the gold is retained there on amalgamated copper plates. Slightly inclined amalgamated plates arranged in steps are placed in front of the battery, and over these the crushed ore pulp passes slowly, the gold being retained by the amalgamated surfaces. The remaining product, or 'tailings', which may still contain some gold, is then treated either by 'concentration' and the concentrates chlorinated, or the whole is treated by the cyanide process.

Gongylite

Gongylite is a yellowish-brown coloured variety of Agalmatolite found in Finland.

Goslarite

Goslarite (white vitriol) is hydrated zinc sulphate. It is a rare mineral precipitated from water seeping through the walls of lead mines and is formed by the decomposition of sphalerite.

Gossan

Gossan is decomposed rock, usually reddish or ferruginous (owing to oxidized pyrites), forming the upper part of a metallic vein.

Granite

Granite is a plutonic, igneous, crystalline, granular rock, consisting generally of quartz, feldspar, and mica, mixed up without regular arrangement of the crystals, and usually of a whitish, greyish, or flesh-red colour. The grains vary in size from that of a pin's head to a mass of almost one meter, but they seldom exceed about two centimetres. When they are of this size, or larger, the granite is said to be 'coarse-grained'.

Granite is one of the most abundant of the igneous rocks seen at or near the surface of the earth, and was formerly considered as the foundation rock of the globe, or that upon which all sedimentary rocks repose; but it is now known to belong to various ages from the Pre-Cambrian to the Tertiary, the Alps of Europe containing granite of the later age. In Alpine situations it presents the appearance of having broken through the more superficial strata; the beds of other rocks in the vicinity rising towards it at increasing angles of elevation as they approach it. It forms some of the most lofty of the mountain chains of the eastern continent, and the central parts of the principal mountain ranges of Scandinavia, the Alps, the Pyrenees, and the Carpathian Mountains are of this rock. Granite supplies one of the most durable materials for building, as many of the ancient Egyptian monuments testify.

Granitel

Granitel is a variety of grey granite with small crystalline granules of feldspar and quartz. Granitel was worked by the Romans as marble.

Granodiorite

Granodiorite is a plutonic igneous rock formed from cooling magma in deep zones of mountain belts. Granodiorite is a course-grained intermediate igneous rock comprised mainly of feldspar and quartz, and either pink or white in colour depending upon the proportions of feldspar and quartz.

Granulite

Granulite is a whitish, granular metamorphic rock, consisting of feldspar and quartz intimately mixed. Granulite is formed deep in the Earth's crust at very high temperature and under high pressures. It is sometimes called whitestone or leptynite.

Graphic Tellurium

Graphic tellurium is an alternative name for Sylvanite.

Graphite

Graphite (also known as Plumbago, Black-lead, and as Wad) is pure carbon with a relative hardness of 2, occurring naturally and often confused with the heavier molybdenite. Graphite occurs not infrequently as a mineral production, and is found in great purity at Borrowdale in Cumberland, and in large quantities in Canada, Sri Lanka, and at Bohemia. Graphite has an iron-grey colour, a metallic lustre, and a granular texture, and is soft and unctuous to the touch. It is formed from organic materials or by the presence of hydrocarbons in a metamorphic region. Graphite may be heated to any extent in close vessels without change and is exceedingly unchangeable in the air and as such has been used in the manufacture of crucibles, as a lubricant when mixed with oils, and as 'lead' for pencils when mixed with clay. Graphite is also used in burnishing iron to protect it from rust, for giving a smooth surface to casting moulds, for coating wax or other impressions of objects designed to be electro-typed, and for counteracting friction between the rubbing surfaces of wood or metal in machinery. Since the start of the 20th century artificial graphite has been produced by the electric furnace.

Gravel

Gravel is a mixture of coarse sand and small water-worn stones. The term may also be applied to small water-worn stones on their own.

Graywacke

Graywacke is a metamorphic sandstone in which grains or fragments of various minerals, as quartz and feldspar, or of rocks, as slate and siliceous clay rocks, are embedded in an indurated matrix, which may be siliceous or argillaceous. The colours are grey, red, blue, or some shade of these. The term, as used by the earlier writers, included all the conglomerates, sandstones, and shales of the older formations, when these had been subjected to considerable change. At first it was nearly synonymous with the Silurian strata, these, especially in Scotland, yielding the only genuine graywacke. The term has been little used since the start of the 20th century.

Greasy

Greasy refers to a lustre of a mineral which appears to be covered with thin coat of oil.

Green Earth

Green earth (also known as mountain green) is a variety of Glauconite, found in cavities in amygdaloid and other eruptive rock, and used as a pigment by artists.

Greenockite

Greenockite, named after Lord Greenock, is the most common mineral containing cadmium but found in few places and usually as an earthy coating on zinc ores. The largest use of cadmium is for electroplating other metals to form chemical resistant coatings. It has the formulae CdS and a relative hardness of 4.

Greensand

Greensand is a variety of sandstone, usually imperfectly consolidated, consisting largely of glauconite, a silicate of iron and potash of a green colour, mixed with sand and a trace of phosphate of lime. Greensand is often called marl, because it is a useful fertilizer. The greensand beds of the American Cretaceous belong mostly to the Upper Cretaceous.

Greisen

Greisen is a crystalline rock consisting of quartz and mica, common in the tin regions of Cornwall and Saxony.

Greystone

Greystone is a greyish or greenish compact rock, composed of feldspar and augite, and allied to basalt.

Groppite

Groppite is a red or brownish-red coloured crystalline mineral with a hardness of 2.5, comprised mainly of aluminium, iron and magnesia found in limestone. Grossularite (grossular) is a translucent garnet of a pale green colour like that of the gooseberry.

Ground Mass

Ground mass is the fine-grained or glassy base of a rock, in which distinct crystals of its constituents are embedded.

Gummite

Gummite is a yellow amorphous mineral, essentially a hydrated oxide of uranium derived from the alteration of uraninite.

Gymnite

Gymnite, so called because it is found in the Bare Hills, Maryland (gymnos being Greek for naked), is a hydrous silicate of magnesia.

Gypsum

Gypsum is a common monoclinic mineral distributed in sedimentary rocks, often as thick beds. Gypsum is usually found under beds of rock salt as it's one of the first minerals to crystallize from evaporated salt waters and is also produced in volcanic areas and in rock veins due to the action of sulphuric acid. Gypsum is used in the production of plaster of Paris. It is chemically a hydrated calcic sulphate and has the formulae $CaSO_4.2H_2O$ and a relative hardness of 2.

Gypsum is found in a compact state as alabaster, or crystallized as selenite, or in the form of a soft chalky stone, which in a very moderate heat gives out its water of crystallization, and becomes a very fine white powder, extensively used under the name of plaster of Paris. This last is the most common, and is found in great masses near Paris, where it forms the hill of Montmartre, near Aix in Provence, and near Burgos in Spain.

Gypsum may be geologically of any age, but occurs abundantly in the more recent sedimentary formations, and is even now forming, either as a deposit from water holding it in solution, or from the decomposition of iron pyrites when the sulphuric acid combines with lime, or from the action of, sulphurous vapours in volcanic regions on calcareous rocks. When gypsum occurs without water it is called anhydrite, but in its most ordinary state it is combined with water.

Gyrolite

Gyrolite is a white or colourless, glassy-looking secondary mineral. A silicate of calcium, gyrolite is formed through the alteration of pre-existent calcium silicates and is found mainly in crevices of basalt or as white spherules on a basalt groundmass, rarely larger than 5 cm in diameter. Gyrolite was first discovered in 1851 and was confirmed as a distinct mineral in 1855. Gyrolite aggregates are often accompanied by apophyllite, quartz and zeolites such as analcime.

Habit

Habit refers to the physical form of a crystal. It's determined by the shape and relative proportions of the crystal faces.

Hade

In geology a hade is the deviation of a fault plane from the vertical. The direction of the hade is the direction toward which the fault plane descends from an intersecting vertical line.

Haidingerite

Haidingerite is a mineral consisting chiefly of the arsenate of lime.

Hair-salt

Hair-salt is a variety of native Epsom salts occurring in silky fibres.

Halides

Halides refers to a group of minerals that are mostly compounds of halogen elements (bromine, chlorine, fluorine, iodine). Ex: halite, fluorite.

Halite

Halite (common salt) is a mineral found in lustreless cubic crystals. It has the formulae NaCl and a relative hardness of 3.

Hallyoysite

Hallyoysite, named after Omalius d'Halloy, is a claylike mineral, occurring in soft, smooth, amorphous masses, of a whitish colour.

Hamilton Period

The Hamilton period is a subdivision of the Devonian system of America. It is so named from Hamilton, Madison County, New York. It includes the Marcellus, Hamilton, and Genesee epochs or groups.

Hard-pan

Hard-pan is the name given to the stratum of sand or gravel, hardened, usually by iron oxide, which is found at a depth of between 25 and 100 cm in arenaceous formations such as the Bagshot sands.

Hardness

Hardness refers to the resistance of a smooth surface of a rock or mineral to scratching. It is determined by the binding force of atoms within the crystal structure and is measured according to Moh's ten division scale of hardness, ranging from 1 the softest, to 10 the hardest: (1) talc; (2) gypsum; (3) calcite; (4) fluorite; (5) apatite; (6) orthoclase; (7) quartz; (8) topaz; (9) corundum; (10) diamond. Materials, according to this arrangement, which are scratched by quartz and are not scratched by felspar are said to have a hardness between 6 and 7.

Harlech Group

The Harlech group is a minor subdivision at the base of the Cambrian system in Wales.

Harmotome

Harmotome or cross-stone is a hydrous silicate of alumina and baryta which occurs in right rectangular prisms terminated by four rhombic planes corresponding to the solid angles of the prism; but more frequently in twin-crystals formed by the intersection of two flattened prisms at right angles to each other. Its prevailing colour is white, but may be yellow, red, or brown; it is translucent or semi-transparent, with a somewhat pearly lustre, and is hard enough to scratch glass having a relative hardness of 5. It occurs mostly in volcanic rocks, especially basalt. A related mineral, called lime harmotome, and Phillipsite, contains lime in place of baryta.

Hastings Sands

Hastings sands is the lower group of the Wealden formation. It is so called from its development around Hastings, in Sussex, England.

Hatchettine

Hatchettine, named after its discoverer, Charles Hatchett is a mineral tallow. It is a waxy or spermaceti-like substance, commonly of a greenish yellow colour.

Hauerite

Hauerite, named after Von Hauer, is a native sulphide of manganese, a reddish brown or brownish black mineral.

Hausmannite

Hausmannite is an ore of manganese. It has the formulae Mn_2O_4 and a relative hardness of 6.

Hauyne

Hauyne has the formulae (Na,Ca)4-8(Al_6Si_6)O_{24}(SO_4,S)1-2. It has a relative hardness of 6. It is of interest to collectors. Occurs in igneous and volcanic rocks.

Heave

In geology, a heave is a horizontal dislocation in a metallic lode, taking place at an intersection with another lode.

Hedenbergite

Hedenbergite is a green or black coloured rock-forming mineral found in contact metamorphic rocks such as skarn, and also in certain igneous rocks and ore bodies. Hedenbergite was named by the Swedish chemist Johan Berzelius after his colleague Ludvig Hedenberg. Hedenbergite is a pyroxene containing iron and calcium. Crystals are rare, but those which do occur appear in short prisms with square cross-sections.

Helminthite

A helminthite is one of the sinuous tracks on the surfaces of many stones, and popularly called worm trails.

Helvine

Helvine is a mineral of a yellowish colour, consisting chiefly of silica, glucina, manganese, and iron, with a little sulphur.

Hemachate

Hemachate is a species of agate, sprinkled with spots of red jasper.

Hematite

Hematite is a name applied to two ores of iron, red hematite and brown hematite. They are both of a fibrous structure, and the fibres, though sometimes nearly parallel, usually diverge or even radiate from a centre. They rarely occur amorphous, but almost always in concretions, reniform, globular, botryoidal, stalactitic, etc. The red hematite is a variety of the red oxide, and is one of the most important iron-ores. The brown hematite is a variety of the brown oxide or hydrate; its streak and powder are always of a brownish yellow. Hematite has been found in enormous deposits in the United States and elsewhere. Many samples are soft as the hematite is sedimentary or weathered iron oxide and the true hardness is difficult to measure. Hematite has the formulae Fe2O3 and a relative hardness of 7.

Hepatite

Hepatite is a variety of barite. It emits a feted odour when rubbed or heated.

Herderite

Herderite, named after its discoverer Baron von Herder, is a rare fluophosphate of glucina, occurring in small white crystals.

Heulandite

Heulandite, named after the English mineralogist Heuland, is a mineral of the Zeolite family, often occurring in amygdaloid, in foliated masses, and also in monoclinic crystals with pearly lustre on the cleavage face.

Hiddenite

Hiddenite, named after W. E. Hidden, is an emerald-green variety of spodumene found in North Carolina and used a gem stone.

Hisingerite

Hisingerite, named after the Swedish mineralogist W. Hisinger, is a soft black, iron ore, nearly earthy, a hydrous silicate of iron.

Hitch

In geology a hitch is a small dislocation of a bed or vein.

Holosiderite

Holosiderite is a meteorite consisting of metallic iron without stony matter.

Homilite

Homilite is a borosilicate of iron and lime, near datolite in form and composition.

Hone

Hone is the name given to several varieties of slaty stones employed in whetting knives, razors, or other edge-tools. They are usually pieces of hard close-grained clay-slate, containing minute particles of quartz, with a uniform consistence. The best-known varieties are the Ayr stone, so called from being found in the river Ayr, in Scotland; the Charnley Forest stone, found in Charnwood Forest, Leicestershire; the German hone, the Canada oil-stone, Turkey oil-stone, etc.

Hopeite

Hopeite, named after Professor Hope, of Edinburgh, is a hydrous phosphate of zinc occurring in transparent prismatic crystals.

Horn Mercury

Horn mercury is a mineral chloride of mercury, having a semi translucent, hornlike appearance.

Horn Quicksilver

Horn quicksilver is a native calomel, or bichloride of mercury.

Horn Silver

Horn Silver is a native chloride of silver, so called because when fused it assumes a horny appearance.

Hornblende

Hornblende is a common member of the amphibole group. It can be told from pyroxene by cleavage and is found in igneous and metamorphic rocks, including often granite.

Hornfels

Hornfels is a fine-grained metamorphic rock formed mainly from sedimentary rocks and volcanic tuff in zones of contact metamorphism surrounding igneous intrusions, especially granitic masses, at high temperature and extremely high pressures often at great depths within the Earth's crust. Various varieties of hornfels occur, chiastolite hornfels are grey or brown in colour and contain mica and quartz together with andalusite and cordierite. Garnet hornfels is generally dark coloured with reddish patches and contains crystals of garnet, mica, quartz and feldspar.

Hornito

A hornito is a low, oven-shaped mound, common in volcanic regions, and emitting smoke and vapours from its sides and summit.

Hortonite

Hortonite, named after a Mr Horton, is a variety of pyroxene, of which it is a steatic pseudomorph. Hortonite is found in Orange County, New York with chrondrodite.

Houghite

Houghite, named after Dr Franklin B Hough of Somerville, New York, is a variety of Hydrotalcite found in St Lawrence county, New York.

Hubnerite

Hubnerite is a yellow to reddish brown mineral found in metal-bearing veins. Hubnerite was identified in 1865 and named after Adolph Hubner, a German metallurgist, who first analysed it. Hubnerite is a member of the wolframite series, the mineral gradually has its manganese atoms replaced by iron until it becomes wolframite. Hubnerite has a relative hardness of 4.5.

Humite

Humite, named after the English mineralogist Sir Abraham Hume, is a fairly rare mineral of a transparent vitreous brown colour, found in the ejected masses of Vesuvius and in hydrothermal veins. It is a silicate of iron and magnesia, containing fluorine. Humite was confirmed as a distinct species of mineral in 1813.

Huronian-rocks

In geology, huronian-rocks is a term applied to certain rocks on the banks of Lake Huron, consisting of quartzite, with masses of chloritic schist. They occupy the same relative position as the upper parts of the Archaean rocks of Britain.

Hyacinth

Hyacinth or jacinth is a red variety of zircon, sometimes used as a gem stone. Its crystals, when distinct, have the form of a four-sided prism, terminated by four rhombic planes, which stand on the lateral edges. Its prevailing colour is a red, more or less tinged with yellow or brown. The name hyacinth is also given to varieties of the garnet or cinnamon stone, the sapphire, and topaz.

Hyalite

Hyalite (Muller's glass) is a clear, glassy variety of opal, never opalescent though often faintly iridescent.

Hyalophane

Hyalophane is a species of the feldspar group containing barium.

Hydromagnesite

Hydromagnesite is a hydrous carbonate of magnesia occurring in white, earthy, amorphous masses.

Hydromica

Hydromica is a variety of potash mica containing water. It is less elastic than ordinary muscovite.

Hydromica Schist

A hydromica schist is a mica schist characterized by the presence of hydromica. It often has a silky lustre and almost soapy feel.

Hydrophane

Hydrophane is a porous, translucent opal, which on being placed in water becomes nearly transparent as the water is absorbed into the pores.

Hydrotalcite

Hydrotalcite is a hexagonal, massive, foliated, fibrous white translucent mineral of pearly lustre and a greasy feel composed of aluminium, magnesia and water. It is found in serpentine.

Hydrothermal

Hydrothermal refers to the alteration of minerals or rocks by super-heated mineral rich fluids, usually water, within a crystallizing magma.

Hydrozincite

Hydrozincite (zinc bloom) is a hydrous carbonate of zinc, usually occurring in white earthy incrustations mainly in deposits associated with smithsonite and occurs as a result of the oxidation of zinc bearing minerals. A major ore of zinc if found in economic quantities.

Hypersthene

Hypersthene or Labrador hornblende is an orthorhombic mineral of the pyroxene group, of a greyish or greenish black colour, often with a peculiar bronze like lustre (schiller) on the cleavage surface. It was first found on the coast of Labrador, and was called Labrador hornblende.

Hypogene

In geology, the term hypogene describes rocks formed or crystallized at depths beneath the earth's surface. It is said of granite, gneiss, and other rocks, whose crystallization is believed of have taken place beneath a great thickness of overlying rocks.

Ice Spar

Ice spar is a variety of feldspar, the crystals of which are very clear like ice.

Iceland Spar

Iceland spar is a transparent variety of calcite, the best of which is obtained in Iceland. It is used for the prisms of the polariscope, because of its strong double refraction.

Idrialine

Idrialine is a bituminous substance obtained from the mercury mines of Idria, where it occurs mixed with cinnabar.

Igneous Intrusion

Igneous intrusion refers to a body of igneous rock that has made its way into pre-existing rock (known as country rock). Igneous intrusions are emplaced as magma, which is less dense than solid rock and therefore tends to move upwards. It can then force its way through cracks in the rocks and can wedge them apart or, if it is hot enough, it can melt and replace them.

Igneous intrusions can be of a variety of shapes and sizes, ranging from huge batholiths to bodies only one or two meters across. The general term 'pluton' can be applied to any of these. Intrusions may cut across the bedding of the country rock. They are then termed 'discordant' or 'transgressive'; dykes and laccoliths are examples. Intrusions that follow the bedding of the country rock, such as sills, are termed 'concordant'.

Igneous Rock

Igneous rock is rock that has originated from molten or semi-molten magma. Igneous rocks are composed almost entirely of silicate minerals. Of many different types, they can be classified in various ways: by composition, crystal size, or mode of occurrence. One method is to divide them according to their silica content into three main groups: acid rocks, basic rocks, and intermediate rocks. There is, however, continuous gradation from acid to basic. Acid rocks are characterized by the presence of 10 per cent or more of quartz, usually accompanied by lighter- coloured minerals such as orthoclase feldspar and muscovite. This acid group includes rocks such as granite and rhyolite. Molten acid lava is very viscous, and its viscosity prevents the molten lava flowing very far and produces tall, conical volcanoes of the Vesuvius type. Intermediate rocks have less than 10 per cent quartz; diorite and andesite are typical of this group.

Ilmenite

Ilmenite or titaniferous iron oxide is an iron-black or steel-grey coloured common rock-forming mineral which is chiefly found in small crystals in basic igneous rocks and schists, consisting chiefly of the oxides of iron and titanium. It has weak magnetism and a relative hardness of six. It is the major ore of titanium. It is also known as menaccanite after Menaccan, in Cornwall, where it was first found.

Ilvaite

Ilvaite (yenite or jenite) is a silicate of iron and lime occurring in black prismatic crystals and columnar masses.

Mineralogy

In mineralogy, an inclusion is a foreign substance, either liquid or solid, usually of minute size, enclosed in the mass of a mineral.

Indicolite

Indicolite is an indigo-blue coloured variety of tourmaline.

Indium

Indium is a soft, rare metal first discovered by Reich and Richter in 1863 by means of spectroscopic analysis in the zinc-blende of Freiburg. Indium has a silver-white colour, is soft, and marks paper like lead. The metal is related to aluminium, and its spectrum exhibits two characteristic lines, one violet and another blue.

Intermediate Rock

Intermediate rock refers to an igneous rock that is transitional between acidic and basic rocks. Have a silica content between 54-65%.

Intratelluric

In geology, the term intratelluric refers to something situated, occurring, or originating in the interior of the earth.

Iodyrite

Iodyrite is a yellowish coloured mineral form of Silver iodide.

Iron

Iron is the commonest and best-known metal. It has the symbol Fe (from the Latin ferrum). Iron is one of the most tenacious of the metals, having a breaking strain of 106,000 lbs. per square inch of section; and two pieces can be perfectly welded together when raised to a white heat. It is so ductile that it can be drawn into wire as fine as the human hair.

1. *Magnetic Iron Ore*. This, the richest of all the ores of iron, and contains, when pure, 72.41 per cent of metallic iron. It is iron-black in colour, with a metallic lustre, highly magnetic (especially the specimens of it that are called native loadstone), and extremely infusible. Some mountains in Lapland and Chilli consist almost entirely of this variety of ironstone. In Sweden it exists in great abundance and purity, and the bar-iron produced from it is much in demand by steel manufacturers. It is plentiful also in Norway, Russia, the East Indies, China, and the north-east part of the United States. Various parts of Great Britain also possess deposits of magnetic ore. Its specific gravity varies from 4.24 to 5.4
2. *Haematite or Specular Iron Ore*, Red Haematite, or ferric oxide. This mineral in its purest state contains about 70 per cent of iron. Specular ore is a deep steel-grey in colour, with a brilliant, and often iridescent tarnish externally; its fracture exhibits a brilliant lustre. It is opaque in large fragments, but the edges of small

thin scales are of a blood-red colour by transmitted light. This ore is found in the older rocks, especially gneiss and granite, both in beds and veins. Great Britain has rich deposits of haematite in Cumberland, Lancashire, Devon and Cornwall, the red ores being chiefly utilized by British smelters. France, Germany, Russia, and North and South America have large deposits of the crystalline variety.

3. ***Brown Iron Ore,*** Brown Haematite. This variety consists essentially of hydrated ferric oxide, and contains when pure about 60 per cent of iron, along with about 16 per cent of combined water. Brown iron ore occurs plentifully in France, Germany, Belgium, and in England, chiefly in the Forest of Dean, in Devon, Lincolnshire, and near Durham. Brown haematite is generally a yellow powder, sometimes passing into a brown or velvet black. It affords a very malleable and much harder iron than the red ore, and very good steel. Before the blowpipe it blackens and magnetizes, but after calcination and cooling the powder becomes red, and in this state is much used for polishing metals. Bog iron ore is a variety of brown haematite which occurs in most European countries, and is so named from its being chiefly found in marshy places. It is considered to be of recent formation, and the iron obtained from it can but rarely be used for sheet-iron, and never for wire.

4. ***Spathic Iron Ore.*** This mineral, as the name implies, resembles rather an earthy than a metallic substance. It consists essentially of ferrous carbonate, and when pure contains 48.27 per cent of iron. It occurs in the older rocks in veins and beds, among the chief deposits being in Styria and Westphalia, the Pyrenees, Colombia, and Great Britain. This ore is very valuable for making steel, being free from phosphorus, but containing manganese. Spathic ironstone is often associated with considerable

quantities of clayey and coaly matter; when the former substance predominates the ore is known as argillaceous or clay ironstone; when the coaly matter is in excess the ore is called carbonaceous or black-band ironstone. These varieties occur in most of the coal-fields of Great Britain, and formerly supplied the greater part of the iron produced there. It is also worked in France. The colour of the clayey carbonates of iron varies from reddish-brown through yellow-brown to dark brownish black.

5. *Iron Pyrites,* This mineral, when pure, consists of 53.33 per cent of iron combined with 46.67 per cent of sulphur, and is the most widely distributed of all the ores of iron. It occurs in many forms disseminated in rocks, veins, and beds, and is frequently found in coal-seams. The United States have immense deposits of it. The ordinary colour is brass-yellow. Before the blowpipe it melts, giving out a sulphurous odour, and leaving a blackish slag, which is magnetic. This mineral is chiefly used as a source of sulphur in sulphuric acid factories, and not as a source of iron. In Siberia it is worked for the small percentage of gold it contains.

Iron Clay

Iron clay is a yellowish clay containing a large proportion of an ore of iron.

Iron Flint

Iron flint is an opaque, flint-like, ferruginous variety of quartz.

Ironstone

Ironstone is a rock containing at least fifteen percent iron. Ironstone forms in igneous rocks through mahmatic differentiation - dense heavy minerals such as magnetite crystallise first and sink to the base of the molten magma to form an iron-rich layer. Ironstone also forms in sedimentary rocks such as limestone and sandstone.

Isinglass

Isinglass is a popular name for mica, especially when in thin sheets.

Isometric

Isometric refers to cubic. Three axis, all the same length and at right angles to each other.

Itacolumite

Itacolumite is a laminated, light-coloured, granular, siliceous rock, consisting of fine quartz grains, and flakes of mica, talc, and chlorite, often occurring in regions where gold and diamonds are found.

Mesonite, named after Professor Jameson, of Edinburgh, is a steel-grey mineral, of metallic lustre, commonly fibrous massive. It is a sulphide of antimony and lead, with a little iron.

Jargon

Jargon (also known as Jargoon or Jacynth or Matara Diamond) is a colourless, yellowish or smoky coloured variety of zircon found in Sri Lanka.

Jarosite

Jarosite, named after Barranco Jaroso in Spain, is an ochre-yellow mineral occurring in minute rhombohedral crystals. It is a hydrous sulphate of iron and potash.

Jasper

Jasper is an impure, opaque, cryptocrystalline variety of quartz that takes a high polish and was formerly used for vases, seals, snuffboxes, etc and is now used as a gemstone.

Jasper is less hard than flint and common quartz and is usually stained by impurities and occurs in various colours, such as red, green, yellow, and blue. When the colours are arranged in bands, the mineral is called riband jasper; a variety containing alternating bands of red and green is known as Siberian jasper. Mottled yellow or brown varieties of jasper are called Egyptian jasper. Agate jasper is intermediate in structure between true jasper and chalcedony. Inclusions of red jasper occur in heliotrope. Jasper is found in metamorphic rocks, and often occurs in very large masses.

Jasper Opal

The jasper opal is a yellow variety of opal resembling jasper.

Jaspilite

Jaspilite is a compact siliceous rock resembling jasper.

Jasponyx

Jasponyx is an onyx, part or all of whose layers consist of jasper.

Jeffersonite

Jeffersonite, named after Thomas Jefferson, is a variety of pyroxene of olive-green colour passing into brown. It contains zinc.

Jet

Jet is a variety of lignite, of a very compact texture and velvet black colour, susceptible of a good polish, glossy in its fracture, which is conchoidal or undulating and often wrought into mourning jewellery, toys, buttons, etc. It was formerly also called black amber. In Great Britain jet is found chiefly at Whitby and the surrounding coast of North Yorkshire in beds of the Upper Lias shale.

Jurassic

The Jurassic was the second geological period of the Mesozoic era. It followed the Triassic, which ended about 190 million years ago, and extended until the beginning of the Cretaceous period, about 139 million years ago. It was named in 1829 by A Brongniart after the Jura Mountains on the borders of France and Switzerland.

Kabook

Kabook is a clay ironstone found in Sri Lanka.

Kainite

Kainite is a compound salt consisting chiefly of potassium chloride and magnesium sulphate, occurring at the Stassfurt salt mines in Prussian Saxony. Kainite dissolves easily in water and has a relative hardness of 3. Kainite is valuable for the production of double sulphate of potash and magnesia, and is used as a manure.

Kalicine

Kalicine is native potassium bicarbonate occurring as a white crystalline mineral.

Kalsilite

Kalsilite is a rare hexagonal silicate of potassium and aluminium found in some lavas.

Kamacite

Kamacite is an alloy of iron and nickel occurring in meteorites as bar-shaped masses.

Kampylite

Kampylite or campylite is a variety of mimetite or arsenate of lead in hexagonal prisms of a fine orange yellow.

Kaolinite

Kaolinite is a hydrous silicate of aluminium and a relative hardness of 3. It has a crumbly habit and forms the basis of most clay. Kaolinite is derived from the decomposition of feldspars, particularly aluminium silicates and is one of the most important of the natural industrial substances, used for bricks, ceramics, and many other applications.

Kapia

Kapia is the fossil resin of the kauri tree of New Zealand.

Karpholite

Karpholite is a fibrous mineral occurring in tufts of a straw-yellow colour. It is a hydrous silicate of alumina and manganese.

Keilhauite

Keilhauite is a mineral of a brownish black colour, related to titanite in form. It consists chiefly of silica, titanium dioxide, lime, and yttria.

Kermes Mineral

Kermes mineral is an amorphous antimony trisulphide, a brown-red powder used in the preparation of artists' colours.

Kernite

Kernite is a major ore of boron. It is formed in playa lakes with the boron supplied by thermal springs passing through the underlying volcanic rock. It has the formulae $Na_2B_4O_7$ù$_4H_2O$ and a relative hardness of 3.

Kidney Ore

Kidney ore is a variety of hematite or iron sesquioxide, occurring in compact kidney-shaped masses.

Kiefekil

Kiefekil is a species of clay.

Kimberlite

Kimberlite is an igneous rock containing very little silica.

Kimmeridge Clay

Kimmeridge clay is a bluish slaty clay, containing some carbonate and sulphate of lime, found in thick deposits in the south of England at and around Kimmeridge in Dorset, and in the north of France. It is a member of the Upper Oolite.

Knebelite

Knebelite, named after Major von Knebel, is a mineral of a grey, red, brown, or green colour, and glistening lustre. It is a silicate of iron and manganese.

Kobellite

Kobellite, named after Franz von Kobell, of Munich, is a blackish grey mineral, a sulphide of antimony, bismuth, and lead.

Kottigite

Kottigite is a dark red or brown coloured mineral formed as a weathering product of zinc minerals such as sphalerite. It is a minor ore of zinc and is named after Otto Kottig, a chemist who discovered it in Germany, kottigite being confirmed as a distinct species of mineral in 1850.

Kunzite

Kunzite is a transparent form of spodumene. It was discovered in 1902 by George Kunz, the German mineralogist, near the Pala Chief Mountain in California.

Kupletskite

Kupletskite is a mineral confirmed as a distinct species in 1956 following work carried out by Boris Kupletski and his wife in Russia into the properties of astrophyllite. Astrophyllite undergoes a reaction in which the titanium is replaced by niobium, the resulting mineral being kupletskite which unlike astrophyllite has a triclinic system of symmetry.

Kutnohorite

Kutnohorite is a rare greenish grey or pinkish coloured carbonate mineral confirmed as a distinct species of mineral in 1901 and named after the place where it was discovered - Kutna Hora in the Czech Republic. Kutnohorite is a carbonate of calcium. Iron and manganese related to ankerite and dolomite. Kutnohorite has a relative hardness of 3.5 to 4.

Labradorite

Labradorite (Labrador feldspar) is a mineral found on the coast of Labrador, and formerly called Labrador hornblende, though that is the designation of hypersthene. It is one of the plagioclase feldspars, also known as soda-lime feldspars.

Laccolith

A laccolith (or laccolite) is a dome of igneous rock between two layers of older sedimentary rock. They are formed by the intrusion of magma, which forces the overlying strata into a dome.

Lamellar

Lamellar refers to composed of thin layers, scales, or plates

Lamination

In geology, lamination is the arrangement of rocks in thin layers or laminae. This arrangement prevails amongst all the varieties of gneiss, mica schist, chlorite schist, hornblende schist, etc.

Lanarkite

Lanarkite is a mineral consisting of sulphate of lead, occurring either massive or in long slender prisms, of a greenish-white or grey colour.

Lanthanite

Lanthanite is hydrous carbonate of lanthanum, found in tabular white crystals.

Lapis Lazuli

Lapis lazuli is a blue aluminious mineral formerly used in the manufacture of ultramarine pigment. Lapis luzuli has a vitreous lustre and an uneven fracture. It scratches glass, is opaque, easily broken and has a specific gravity of 2.45.

Laterite

Laterite (named from the Latin later, meaning a brick or tile), is a red or brick-like mineral substance, sometimes forming a clay, found in various regions, especially in the Deccan and other parts of India, and in Sri Lanka.

Laumontite

Laumontite, named after its discoverer, Dr. Laumont, is a form of anaclime (or analcite) with a powdery white surface. It is characterised by the fact that it is monoclinic. Exposed to the air, it loses water, becomes opaque, and crumbles.

Laurentian

In geology, Laurentian is a term applied to a vast series of stratified and crystalline rocks of gneiss, mica-schist, quartzite, serpentine, and limestone, about 40,000 feet in thickness, lying northward of the St Lawrence in Canada.

The Laurentian is the lowest fossiliferous system of rocks, if its characteristic and only fossil, the Eozoon canadense, can be ranked as a fossil.

Laurite

Laurite is a rare sulphide of osmium and ruthenium found with platinum in Borneo and Oregon.

Lava

Lava is the general term for all rock-matter that flows, or has flowed, in a molten state from volcanoes, and which when cooled down forms varieties of tufa, trachyte, trachytic greenstone, and basalt, according to the varying proportions of felspar, hornblende, augite, etc, which enter into the composition of the mass, and according to the slowness or rapidity with which it has cooled. The more rapidly this process of cooling goes on the more compact is the rock. Lava beds are of two kinds, namely, contemporaneous and intrusive. A contemporaneous lava bed is one which has been poured out over the surface of one deposit, and covered by subsequent deposits. Such a bed is in its natural position, and usually alters only the bed beneath it. Intrusive beds are those which have been forced up in a molten state through or between strata, altering those on both sides.

Lava Millstone

Lava Millstone is a hard and coarse basaltic millstone, obtained from quarries near Andernach on the Rhine.

Lazulite

Lazulite or blue-spar is a rare mineral and a minor gem stone of a light indigo-blue colour, occurring in small masses, or in monoclinic crystals. It is a hydrous phosphate of alumina and magnesia and iron. It is usually found in quartzites associated with kyanite, andalusite, corundum and rutile.

Lazurite

Lazurite a relative hardness of 5. It is usually a deep blue colour, the greenish variety is called lapis lazuli and is very rare. Lazurite is found in metamorphic limestones associated with calcite and pyrite.

Leadhillite

Leadhillite, so called from having been first found at Leadhills, Scotland, is a transparent to translucent mineral of a yellowish or greenish white colour, consisting of the sulphate and carbonate of lead. It is a secondary mineral confirmed as a distinct species in 1832 and formed in the oxidation zones of lead deposits where it is commonly associated with anglesite, cerussite, chalcosite, dioptase, galena, linarite, malachite, silver, willemite and wulfenite.

Lenticular

Lenticular refers to having a lens-like shape.

Lepidocrocite

Lepidocrocite is a secondary mineral that is generally found in association with goethite and other iron-bearing minerals, especially hematite, ilmenite, limonite, magnetite, pyrite and siderite. Lepidocrocite is very similar to goethite, the two having the same chemical formula but differing in specific gravity - due to the different manner in which they formed - lepidocrocite being heavier.

Lepidolite

Lepidolite is a relatively rare species of mica, of a lilac or rose-violet colour, found in pegmatic dikes and usually associated with lithium bearing minerals such as pink and green tourmaline. Often found intergrown with muscovite. Used as a source of lithium and it's used in the manufacture of heat resistant glass.

Lepidomelane

Lepidomelane is an iron-potash mica, of a raven-black colour, usually found in granitic rocks in small six- sided tables, or as an aggregation of minute opaque scales.

Leucite

Leucite is a rather rare mineral occurring only in igneous rocks and usually in recent lavas. It is found in rocks where the amount of silica in the magma was not sufficient to form feldspar. It is not found, therefore, in rocks that contain quartz. It's name is derived from the Greek word for ' white'. It has the formulae $KAlSi_2O_6$ and a relative hardness of 6.

Leucophane

Leucophane or leucophanite is a mineral of a greenish yellow colour. It is a silicate of glucina, lime, and soda with fluorine.

Leucopyrite

Leucopyrite is a mineral of a colour between white and steel-grey, with a metallic lustre, and consisting chiefly of arsenic and iron.

Leucoxene

Leucoxene is a nearly opaque white mineral, in part identical with titanite, observed in some igneous rocks as the result of the alteration of titanic iron.

Levyne

Levyne, named after the English mineralogist Levy, is a whitish, reddish, or yellowish, transparent or translucent mineral, allied to chabazite.

Lherzolite

Lherzolite, named after Lherz in the Pyrenees is an igneous rock consisting largely of chrysolite, with pyroxene and picotite (a variety of spinel containing chromium).

Libethenite

Libethenite, named after Libethen, in Hungary, where it was first found, is a rare secondary mineral of an olive-green or deep-green colour, commonly occurring in orthorhombic crystals.

Lignite

Lignite (brown coal, wood coal) is a mineral coal closely related to jet, retaining the texture of the wood from which it was formed, and burning with an empyreumatic odour. It is of more recent origin than the anthracite and bituminous coal of the proper coal series, and often contains untransformed woody material.

Limestone

Limestone is one of the various forms in which calcic carbonate (carbonate of lime) occurs naturally. Other natural forms of the carbonate are marble, chalk, and calc-spar. When pure it has the composition of 57 per cent of lime and 43 of carbon dioxide. Most varieties of limestone are highly impure, and contain magnesium carbonate, silica, alumina, and iron compounds. They are comparatively soft and can be scratched with a knife, and all effervesce when a drop of dilute acid or vinegar is added.

Limonite

Limonite is the hydrated oxide of iron. It never occurs crystallised, but in fibrous, earthy, stalactitic mammillary, porous, or concretionary masses, and often as pseudomorphs formed by the weathering of other minerals such as pyrites or marcasite. It is black or brown in colour and the streak or fine powder is yellow, distinguishing it from haematite. It has a relative hardness of 5. 5. Impure limonite is found in moorland clay soils and also in meadows and bogs in spongy nodules, and is hence known as bog iron ore. In fresh-water lakes it is often deposited as a brownish slime by the action of plants on the ferrous carbonate and in Sweden and Norway this lake ore is periodically collected by raking the bottom of the shallow pools. Earthy limonite, or limonite mixed with clay is known as yellow ochre.

Linarite

Linarite is a hydrous sulphate of lead and copper occurring in bright blue monoclinic crystals. Linarite is a rare secondary mineral formed in lead and copper veins that have been altered by the action of water and other circulating fluids. It is most often associated with anglesite, brochantite, cerussite, chalcanthite, chalcopyrite, galena and malachite. Linarite is named after the place where it was first discovered, Linares in Spain. It was confirmed as a distinct species in 1839. It has a relative hardness of 2.5 and is fairly heavy.

Liroconite

Liroconite is a hydrated arsenate of copper, occurring in obtuse pyramidal crystals of a sky-blue or verdigris-green colour.

Litharge

Litharge is a red lead oxide mineral, dimorphous with massicot, and is often found as a crust enveloping an inner core of massicot in a rock base. Linarite is amphoteric - it dissolves equally well in acids and alkalis, and was formerly used to prepare lead salts and plumbates to make rust-resistant paints.

Lithia

Lithia is an oxide of the metal lithium, which was at first found in a mineral called petalite. It is of a white colour, very soluble in water, acrid, caustic, and acts on colours like other alkalies.

Lithiophilite

Lithiophilite is a phosphate of manganese and lithium; a variety of triphylite.

Lithographic Limestone

Lithographic limestone is a compact, fine-grained limestone, obtained largely from the Lias and Oolite, especially of Bavaria, and extensively used in lithography.

Lithomarge

Lithomarge is a compact variety of kaolin or of clay, soft, unctuous, and friable. It is usually white or grey and is particularly found in Cornwall, Germany and America.

Lithophyse

Lithophyse is a spherulitic cavity often with concentric chambers, observed in some volcanic rocks, as in rhyolitic lavas. It is supposed to be produced by expanding gas, hence the name.

Lizard Stone

Lizard stone is a kind of serpentine marble from near Lizard Point, Cornwall, England, used for ornamental purposes.

Loam

Loam is a soil compounded of various earths, of which the chief are sand, clay, and carbonate of lime or chalk, the clay predominating. Decayed vegetable and animal matter, in the form of humus, is often found in loams in considerable quantities, and the soil is fertile in proportion.

London-clay

London-clay is the most important of the Eocene tertiary formations of Great Britain, largely developed in the valley of the Thames under and around the metropolis. This formation consists of a bluish or brownish clay containing layers of argillaceous nodular limestone. The shells, fruits, etc, found in the London-clay mostly belong to genera now inhabiting warmer seas than those of Britain.

Lucullite

Lucullite, named after Lucullus, a Roman consul, famous for his great wealth and Luxury, is a variety of black limestone, often polished for ornamental purposes.

Ludlamite

Ludlamite is a mineral occurring in small, green, transparent, monoclinic crystals. It is a hydrous phosphate of iron.

Ludlow Rocks

In geology, the Ludlow rocks are a portion of the upper Silurian rocks, characteristically developed at Ludlow in Shropshire.

Ludwigite

Ludwigite, named after the chemist Ernst Ludwig, is a rare mineral, a borate of iron and magnesia, occurring in fibrous masses of a blackish-green colour.

Lustre

Lustre refers to the general look of a mineral in reflected light. Minerals are divided into two types: metallic and non-metallic. There's no clear dividing line between the two. In general, non- metallic minerals will transmit light through a thin edge, are light coloured, and will have a light or colourless streak. Non- metallic minerals are further described as: vitreous, resinous, pearly, greasy, silky, adamantine.

Luxullianite

Luxullianite is a kind of granite from Luxullian, Cornwall, characterized by the presence of radiating groups of minute tourmaline crystals.

Macrodiagonal

In crystallography, the term macrodiagonal refers to the longer of two diagonals, as of a rhombic prism.

Macrodome

In crystallography, the term macrodome refers to a dome parallel to the longer lateral axis of an orthorhombic crystal.

Made

In mineralogy, made is a term applied to twin-crystals, which are united in various ways. Made is also used as a name for chiastolite or cross-stone.

Magma

Magma refers to molten rock beneath the earth's crust. It solidifies to form igneous rocks.

Magnesian Limestone

Magnesian limestone is a yellowish rock composed of carbonates of lime and magnesia, the latter amounting in some cases to nearly a half. There are several varieties, more or less useful for building or ornamental purposes, which are included under the generic name dolomite. The same name is also given to the whole Permian formation, from this rock being very largely developed in it.

Magnesite

Magnesite is native magnesium carbonate occurring in white compact or granular masses, and also in rhombohedral crystals. It has been used as an ore of metallic magnesium but the primary source of magnesium is sea water. Is a source of magnesia for industrial chemicals. Commonly found in veins and masses derived from the alteration of serpentine through the action of waters containing carbonic acid. Magnesite has the formulae MgCo3 and a relative hardness of 4.

Magnetite

Magnetite is an oxide of iron occurring in isometric crystals, also massive, of a black colour and metallic lustre. It is readily attracted by a magnet and sometimes possesses polarity, being then called loadstone. It is an important iron ore and is also called magnetic iron. Magnetite has the formulae Fe_3O_4 and a relative hardness of 7.

Malachite

Malachite is a native hydrous carbonate of copper, usually occurring in green mammillary masses with concentric fibrous structure. It is a widely distributed copper ore. Found in the oxidized portions of copper veins and is often associated with cuprite, native copper, iron oxides, and sulphides of copper and iron. It often occurs in copper veins that are found in limestone. Green malachite, or malachite proper, admits of a high polish, and is sometimes used for ornamental work. Blue malachite, or azurite, is a related species of a deep blue colour.

Malacolite

Malacolite is a variety of pyroxene.

Manganite

Manganite (grey manganese ore) is a strongly magnetic ore of manganese used in the manufacture of glass, but seldom found in commercial quantities. It is often found in veins associated with granitic igneous rocks. It alters to pyrolusite. Barite and calcite are frequent associates.

Marble

Marble is the name given to certain varieties of limestone capable of receiving a brilliant polish, and which, both from their durability and the beauty of the tints of many of them, have at all periods of the world been greatly in request for purposes of art or ornament. White statuary marble is a pure carbonate of calcium. Marbles have been divided into seven varieties or classes: (1) marbles of a uniform colour, comprehending solely those which are either white or black; (2) variegated marbles, or those in which the spots and veins are interlaced and disposed without regularity; (3) shell marbles, or those which are in part made up of shells; (4) lumachelli marbles, or those which are apparently wholly formed of shells; (5) cipolino marbles, or those which are veined with green talc; (6) breccia marbles, or those which are formed of angular fragments of different marbles united by a cement of some different colour; (7) pudding-stone marbles, or those which are formed of reunited fragments, like the breccia marbles, only with the difference of having the pebbles rounded in place of being angular.

Marcasite

Marcasite is one of the sulphides of iron, a paler form of iron pyrites. It often shows a white colour on a fresh surface. It is found in iron metallic veins and frequently with lead and zinc ores, and is also found in sedimentary rocks. Marcasite most frequently occurs as a replacement deposit in limestone and often in concretions imbedded in clays, marls, and shales. It has a relative hardness of 7.

Margarite

Margarite is a metamorphic mineral related to the micas, but low in silica and yielding brittle folia with pearly lustre. It is associated with staurolite and tourmaline and has a relative hardness of 5.

Margarodite

Margarodite is a hydrous potash mica related to muscovite.

Marl

Marl is an earthy substance essentially composed of carbonate of lime and clay in various proportions. In some marls the argillaceous ingredient is comparatively small, while in others it abounds, and furnishes the predominant characters. The most general use of marl is to improve soils. The fertility of any soil depends in a great degree on the suitable proportion of the earths which it contains and whether a calcareous or an argillaceous marl will be more suitable to a given soil may be determined with much probability by its tenacity or looseness, moisture or dryness. The quicker action and greater efficiency of slaked lime resulted in many districts substituting it for marl.

Marmatite

Marmatite is a ferruginous variety of shalerite or zinc blende, nearly black in colour.

Marmolite

Marmolite is a thin, laminated variety of serpentine, usually of a pale green colour.

Martite

Martite is iron sesquioxide in isometric form, probably a pseudomorph after magnetite.

Mascagnin

Mascagnin or mascagnite, named after its discoverer Mascagni, is a native sulphate of ammonia, found in volcanic districts.

Massive

Massive refers to a mineral that does not show any definite external crystal form or consists of poorly defined masses of small crystals.

Matrix

In mining and geology, a matrix is the rock or main substance in which any accidental crystal, mineral, or fossil is embedded.

Meager

In mineralogy the term meager describes a mineral that is dry and harsh to the touch, such as chalk.

Meerschaum

Meerschaum (sepiolite) is a white or yellowish earthy mineral, consisting of hydrous magnesium silicate. It was first found on the sae shore where, resembling petrified sea-froth was given its name - which in German means sea-froth. Meerschaum lathers like soap when first dug, and was used as a soap by the Tartars. It is highly porous, and may float on water. It is very soft, with a relative hardness of 2 to 2.5 and is easily turned into pipes and cigar-holders. Before use the meerschaum is impregnated with oil, tallow or wax.

Meionite

Meionite is a member of the scapolite group of minerals, occurring in glassy crystals on Monte Somma, near Naples.

Melaconite

Melaconite is an earthy black oxide of copper, arising from the decomposition of other ores.

Melanite

Melanite is a lime-iron variety of garnet of a velvet black or greyish black, occurring always in crystals of a dodecahedral form.

Melanochroite

Melanochroite is a chromate of lead mineral of a red, or brownish or yellowish red colour.

Melanterite

Melanterite is a hydrous sulphate of iron of a green colour and vitreous lustre.

Melaphyre

Melaphyre is a compact black or blackish-grey igneous rock, consisting of a matrix of labradorite and augite, in which are embedded crystals of the same minerals, and sometimes uniaxial mica, hornblende, and iron pyrites. It is essentially a basalt.

Melilite

Melilite is a mineral occurring in small yellow crystals, found in the lavas of Vesuvius, and elsewhere.

Mellite

Mellite (honey stone) is a honey coloured mineral, found in brown coal, and partly the result of vegetable decomposition. It is a mellitate of alumina.

Mendipite

Mendipite is a rare white, grey or colourless mineral with a high density but low relative hardness of 2.5 Mendipite is a halide of lead formed around volcanic vents and in hydrothermal veins. It is most commonly found in association with calcite, cerussite, malachite, manganite, pyrolusite and pyromorphite. Most deposits take the form of columnar or fibrous masses, often with radiating external shapes.

Mesolite

Mesolite is a zeolitic mineral, greyish white or yellowish in colour, occurring in delicate groups of crystals, and also fibrous massive. It is a hydrous silicate of alumina, lime, and soda.

Mesozoic Period

Mesozoic period is the term applied by geologists to the geological period between the Palaeozoic and the Cainozoic. It is coextensive with the secondary formations, and includes the rocks of the Triassic, Oolitic, and Cretaceous groups.

Metacinnabarite

Metacinnabarite is sulphide of mercury in isometric form and black in colour.

Metamorphic Rock

Metamorphic rock is rock that has been altered by intense heat, pressure, or both. It may originally have been sedimentary or igneous rock, or even different metamorphic rock. The changes that take place during metamorphism can affect both the structure of the rock and its composition. Very often the rock is re-crystallized. Contact metamorphism is a localized form of metamorphism that is produced by the heat of an igneous intrusion. Limestone may then be altered into marble, and clay into a hard, tough rock (a hornfels). The zone affected in this way is called a metamorphic aureole. Hydrothermal metamorphism, or metasomatism, is produced by hot aqueous fluids emanating from igneous intrusions. China clay is produced in this way from granite. In dynamic metamorphism, or cataclasis, rocks are broken down mechanically by shearing and crushing; mylonite, a fine-grained banded rock, is a typical product. Regional metamorphism takes place on a large scale.

Metaquartzite

Metaquartzite is a metamorphosed sandstone, in which the grains of silica have fused together to form a hard, massive rock. Metaquartzite contains more than 90 percent quartz, which gives the rock a pale, sugary appearance. Metaquartzite is a fine to medium-grained rock formed at great heat under relatively low pressure.

Miargyrite

Miargyrite is a mineral of an iron-black colour, and very sectile, consisting principally of sulphur, antimony, and silver.

Mica

Mica refers to a group of more than thirty silicate minerals having perfect cleavage in one direction and which easily split into thin, elastic, sheets. Micas are closely associated with clay minerals such as gibbsite and kaolinite. Micas are highly resistant to electricity and heat and are widely used in industry as insulators, and are also added to paint and wallpapers to give them a silky lustre. Ground mica serves as a lubricant and absorbent. They were employed in Russia for window panes, and in that state were called muscovy-glass. Mica enters into the composition of the crystalline rocks, as granite, gneiss, mica schists, chlorites, talcose rocks, and occurs in trappean and volcanic products. It is found also in many sedimentary rocks, as shales and sandstones, giving them their laminated texture. In the latter case, it is derived from the disintegration of the crystalline rocks. It is essentially a silicate of alumina, with which are variously combined small proportions of the silicates of potash, soda, lithia, oxide of iron, oxide of manganese, etc; thus several species have been constituted, and varieties receive the names of biotite, lepidolite, muscovite, lepidomelane, steatite, etc.

Mica Schist

Mica schist or mica slate is a metamorphic rock, composed of mica and quartz. It is highly fissile and passes by insensible gradations into clay-slate.

Micaceous Rocks

Micaceous rocks are rocks of which mica is the chief ingredient, as mica slate and clay-slate.

Microcline

Microcline is a mineral with the formulae $KAlSi_3O_8$ and a relative hardness of 7. It has the same composition as orthoclase but is distinguished by triclinic twinning (usually requiring a microscope to detect). If a feldspar is a deep green it is microcline - and sometimes called 'amazon stone' or amazonite after the Amazon river. It is recognisable from plagioclase by a lack of striations.

Microcrystalline

Microcrystalline refers to a rock in which the crystals are too small to be seen without a microscope.

Migmatite

Migmatite is a regional metamorphic rock composed of two pre-existent formations that have combined but may remain distinguishable. A medium to coarse-grained rock, migmatite forms in distinct bands of different colours - mainly green or blacks.

Milarite

Milarite is a rare mineral containing a large number of metals first identified in 1873 and named after Val Milar in Switzerland. Milarite is a hydrated silicate of potassium, calcium, aluminium and beryllium and is a member of the osumilite group. Crystals of milarite are hexagonal and often have six-sided, prismatic shapes, often very small and coloured green or yellow. Milarite is brittle and has no cleavage and has a relative hardness of 5.5 to 6.

Millerite

Millerite has the formulae NiS and a relative hardness of 4. It is the richest ore of nickel but too scattered to be commercially important. Forms at low temperatures often in cavities and as an altered form of other nickel minerals, or as a crystal inclusion in other minerals.

Mimetite

Mimetite is an arsenochloride of lead and a minor ore of lead and has the formulae $Pb_5(AsO_4)3Cl$ and a relative hardness of 4. It is a relatively rare mineral which occurs in the oxidized portions of lead bearing veins. Mimetite occurs in yellow, orange, brown, green and white colours and is related to and resembles pyromorphite. Milarite was identified in 1832 and belongs to the apatite group.

Mineral Caoutchouc

Mineral caoutchouc is a variety of bitumen, intermediate between the harder and softer kinds. It sometimes much resembles india-rubber in its softness and elasticity, and hence its name.

Mineral Structure

The majority of minerals can, under certain conditions, occur in regularly shaped crystals. Crystals vary greatly in shape and size. Some are minute, visible only through a microscope; others may measure several metres. The shape of a crystal can be as thin as a needle (acicular), or columnar, tabular, fibrous or lamellar. A closer look at crystallized minerals shows that their crystals are structurally nearly always the same. It is then possible to conclude that a mineral has a definite crystal structure, which is not accidental but is determined by certain laws. The science which pursues the study of these natural laws and explains and describes crystal structures of different minerals is called crystallography. Natural scientists have studied crystal structure since ancient times, but crystallography only gained recognition as a true science in the 17th and 18th centuries.

Crystallography, as we know it today, is based upon the findings of many other scientific fields, especially mineralogy, chemistry, physics and mathematics. The basic difference between crystalline and non- crystalline matter does not lie only in the regularity of arrangement of the external faces. After all, crystalline grains in rock minerals often have irregular shapes and yet they are crystals. The basic difference is in their internal structure, the arrangement of the molecules, atoms and ions. These tiny particles are chaotically arranged in gases, liquids and non- crystalline solids. In crystals, however, they have a regular, repeating pattern. Some minerals crystallize in a definite characteristic structure. But the structure of many crystallized minerals varies. Calcite crystals, for example, occur in various forms: high or low rhombohedra, columnar, acicular or tabular forms. These shapes share one characteristic, however; they are symmetrical, and their symmetry is in line with the group symmetry of the smallest particles of the substance. According to this symmetry, crystals are divided into seven major groups, called crystal systems: (1) triclinic; (2) monoclinic; (3) orthorhombic; (4) tetragonal; (5) trigonal; (6) hexagonal; (7) cubic. Crystals are classified as belonging to a particular system according to their axes of symmetry which are, basically, imaginary lines passing through the centre of a crystal, and also according to their number and kind. All crystals belonging to an individual system must have a certain characteristic form, which corresponds to the relative symmetry. This means, in fact, that each crystal system has its own individual crystal form. The crystal structure of every mineral is determined mainly by its chemical composition and by physical conditions during its development, especially temperature and pressure. Crystals with completely even faces do not often occur, for the various external conditions make such perfect growth of faces impossible. The size and the development of the individual crystal faces are not so vital for classification as the angle between faces, which is the same for every crystal of the same system.

Minium

Minium is the red oxide of lead, often designated red lead, and formerly commonly used as a pigment for ordinary purposes.

Miocene

Miocene (derived from the Greek meion, less, and kainos, recent), in geology, is the name given by Sir Charles Lyell to a subdivision of the tertiary strata relating to a period of about 25,000,000 years ago. The terms Miocene and Pliocene are comparative, the first meaning less recent and the other more recent. The Miocene strata contain fossil plants and shells which indicate a warm climate.

Mocha Stone

Mocha stone (moss agate) is a pale translucent agate with a branching pattern resembling mosses or fern fronds. They were commonly used in England for brooches, and came chiefly from Arabia.

Mohs

Mohs is a scale of hardness applied to minerals. It ranges from 1 for talc to diamond at 10.

Molasse

Molasse is a soft greenish sandstone which occupies the country between the Alps and the Jura.

Molybdates

Molybdates refers to a group of minerals in which the molybdate radical (MoO_4) is an important constituent. Ex: wulfenite.

Molybdenite

Molybdenite is a grey, metallic-looking mineral with the formulae MoS2 and a relative hardness of 2. It is the main ore of molybdenum and resembles graphite but has a higher specific gravity and a slight blue tint. Molybdenite is formed by hydrothermals in sulphide ore bodies, and also in re-crystallised granitic rocks such as marble, granite and pegmatite. It is frequently found in association with chalcopyrite, fluorite, pyrite, quartz, scheelite, and wolframite. Molybdenite was identified in 1778 by the Swedish chemist Carl Scheele.

Monazite

Monazite has the formulae $(Ca,La,Nd,Th)PO_4$ and a relative hardness of 6. It is the chief ore of thorium and cerium. Thorium is a radioactive element. It is concentrated in sand due to its durability and high specific gravity. Associated with other heavy minerals such as magnetite, rutile, and zircon.

Monoclinic

Monoclinic refers to a crystal with six faces and three axes of unequal length. Two axes are at right angles to each other and the third is inclined to the plane of the other two. A ream of paper with a long edge sloped at an angle is an example.

Montmorillonite

Montmorillonite is a mineral discovered at Montmorillon in France, and confirmed as a distinct species in 1847. Montmorillonite is a form of clay found in many sedimentary rocks especially bentonite and in some metamorphic rocks. It is formed mainly through the alteration of pre-existent feldspar in rocks that are poor in silica. Montmorillonite is a hydrous hydrated silicate of sodium, calcium, aluminium and magnesium used in the petroleum industry as a drilling mud.

Moroxite

Moroxite is the crystallized form of apatite, occurring in crystals of a brownish or greenish-blue colour.

Mottramite

Mottramite is a vanadate mineral usually occurring as a green to black coloured vitreous crust on other rocks and minerals. Mottramite is found in areas where metal-bearing deposits have been oxidised by the action of air and water or just water. Mottramite was identified in 1876 and named after Mottram St Andrew near Manchester, England where it was discovered. A hydrous vanadate of lead, copper and zinc, it is sometimes used as a minor ore of lead and copper.

Mountain-cork

Mountain-cork or Mountain-leather is a white or grey coloured variety of asbestos, so called from its extreme lightness, as it floats in water.

Mountain-limestone

Mountain-limestone is a series of marine limestone strata, whose geological position is immediately below the coal-measures and above the Old Red Sandstone in England and Ireland, the lower Carboniferous or calciferous sandstones in Scotland. It is otherwise termed Carboniferous Limestone.

Mountain-soap

Mountain-soap is a mineral of a pale brownish-black colour, so named from its soapy feel. It occurs in secondary rocks of the trap formation, and is used in crayon-painting.

Mud

In geology, mud is a mixture of clay and sand with organic matter. Mud may be argillaceous, calcareous, or otherwise, according to the most notable ingredient which enters into its composition.

Mudstone

Mudstone was a term originally applied to certain dark-grey fine-grained shales of the Silurian system, but now extended to all similar shales in whatever formation they may occur.

Muschelkalk

Muschelkalk is a compact hard limestone of a greyish colour found in Germany. It is interposed between the Bunter sandstone, on which it rests, and the Keuper variegated marls, which lie over it, and with which at the junction it alternates, thus forming the middle member of the Triassic system as it occurs in Germany. In England the Keuper rests immediately on the Bunter. It abounds in marine organic remains, its chief fossils being encrinites, ammonites, and terebratulae.

Muscovite

Muscovite has extremely perfect cleavage and thin flakes are flexible. Very common. Used for electrical and heat insulation.

Mylonite

Mylonite is a type of metamorphic rock created when pre-existent rocks are crushed, ground or rolled into a new formation along major fault lines. Mylonite is generated by pressure with very little heat. Mylonite rock is ground to dust and streaked out into bands during its creation.

Natrolite

Natrolite is a colourless, white, or yellow zeolite mineral consisting of sodium aluminium silicate in the form of needle-like orthorhombic crystals and a relative hardness of 6. Natrolite is most often found in cavities in basalt. It is formed through the alteration of plagioclase feldspar, nepheline or sodalite.

Natrolite was confirmed as a distinct species in 1803.

Natron

Natron is a whitish or yellow mineral that consists of hydrated sodium carbonate and occurs in saline deposits and salt lakes.

Needle-ore

Needle-ore, acicular bismuth glance is a native sulphide of bismuth, lead, and copper occurring imbedded in quartz in long, thin, steel-grey crystals, marked with vertical striae, and apparently in four or six sided prisms. It consists of lead 35.8, copper 11, bismuth 36.7, and sulphur 16.5, and usually accompanies native gold.

Neocomian

In geology, Neocomian is a term applied to the lower greensand and Wealden.

Neogene

In geology, Neogene is a name given by some geologists to the Pliocene and Miocene tertiaries to distinguish them from the older Eocene strata.

Neozoic

In geology, Neozoic was a name under which Forbes proposed to include all the strata from the beginning of the Trias up to the most recent deposits; the Mesozoic and Cainozoic of other paleontologists. Forbes suggested this classification on the ground that while there ia a widely-marked distinction between Palaeozoic and Mesozoic fossils, there is no essential difference between Mesozoic and Cainozoic.

Nepheline

Nepheline is a mineral that has the formulae $(Na,K)AlSiO_4$ and a relative hardness of 6. It is formed mainly in pegmatite associated with nepheline syenite, but may also occur in gneiss and schist. Nepheline is associated with feldspars, calcite, cancrinite, hornblende and sodalite.

Nephrite

Nephrite is is a tough, compact variety of the mineral tremolite with a hardness of 6 to 6.5. Polished nephrite has an oily lustre. It is found in Alaska, Mexico, New Zealand, Siberia, and Turkestan.

Neptunian Theory

Neptunian Theory is a name given to a geological theory of Werner's, which referred the formation of all rocks and strata to the agency of water as opposed to the plutonic, igneous, or Huttonian theory.

New Red Sandstone

New Red Sandstone is the lowest group of secondary rocks, lying between the Permian below and the Lias above.

Niccolite

Niccolite (nickeline) is an arsenide of nickel mineral with the formulae NiAs and a relative hardness of 6. It is a minor ore of nickel and often has a copper-like colour. It was confirmed as a distinct species in 1832, but was mined in the Middle Ages by miners believing it to be a copper ore.

Nickel-glance

Nickel-glance is a greyish-white, massive, and granular ore of nickel, on the average consisting of 35.5 nickel, 45.2 arsenic, and 19.3 sulphur, part of the nickel being sometimes replaced by iron or cobalt.

Nigrin

Nigrin is an ore of titanium, found in black grains or rolled pieces, containing about 14 per cent of iron. It occurs in Sri Lanka and Transylvania.

Nodular

Nodular refers to appearing as or composed of irregular lumps of rock or a mineral.

Norfolk-crag

In geology, Norfolk-crag is an English tertiary formation belonging to the older Pliocene, resting on the chalk and London-clay. It consists of irregular beds of ferruginous sand-clay, mixed with marine shells and mastodon and elephant remains.

Northern-drift

In geology, the northern-drift is a name formerly given to boulder-clay of the Pleistocene period, when its materials were supposed to have been brought by polar currents from the north.

Nosean

Nosean is a sodium aluminium silicate sulphate mineral with a vitreous lustre discovered by the German mineralogist Karl Nose, after whom it is named, and confirmed as a distinct species in 1815. Nosean is found in extrusive igneous rocks, especially in alkali-rich, silica-poor lavas, and in phonolite. It has a relative hardness of 5 to 5.5.

Oamaru Stone

Oamaru stone is a kind of limestone, of building quality, found at Oamaru on South Island, New Zealand.

Obsidian

Obsidian is an extrusive igneous rock that is composed mainly of the minerals feldspar and quartz. It has been used as a natural form of glass for its cutting properties since prehistoric times - being known as volcanic glass - and today is used as a gemstone.

Old Red Sandstone

Old Red Sandstone is a geological term made popular by the writings of Hugh Miller, and applied by him to the red sandstone which underlies the Carboniferous System, in contradistinction to the New Red Sandstone, which overlies the latter. It is now generally included in the Devonian System.

Oligocene

The Oligocene was the fifteenth geological period, 42,000,000 years ago.

Oligoclase

Oligoclase has the formulae $(Na,Ca)AlSi_3O_8$ and a relative hardness of 7. It is a plagioclase feldspar. It is of interest to petrologists and collectors. Varieties: labradorite, anorthite.

Olivenite

Olivenite is a rare olive-green, brown or yellow coloured secondary mineral found in deeply weathered, highly concentrated copper sulphide ore bodies. It is a hydrous arsenate of copper confirmed as a distinct species in 1820 and sometimes mined as a copper ore. Olivenite has a relative hardness of 3.

Olivine

Olivine is a group of minerals characterised by their green colour, the typical olivine has the formulae $(Mg,Fe)_2SiO_4$ and a relative hardness of 7. It's name is derived from the characteristic olive green colour. A common rock forming mineral. A clear green variety is called peridot and has some uses as a gem. Olivine minerals are hard and resistant to heat, and are used as abrasives and refractory sands.

Onyx

Onyx (named on account it resembles the colour of a fingernail) is a variety of chalcedony, similar to agate but featuring straight, rather than curved bands.

Oolite

Oolite is a species of limestone composed of globules clustered together, commonly without any visible cement or base. They vary in size from that of small pin-heads to that of peas. When the grains are very distinct and well-rounded it is called roe-stone; when they are large and pea-like the rock is known as pisolite, pea-grit, or pea-stone. What is known as the Oolite or Oolitic series of rocks in geology, consist of a series of strata comprehending the whole of those peculiar limestones, calcareous sandstones, marls, etc, which underlie the chalk formation and rest on the Lias. It yields in England a vast quantity of excellent freestone and ironstone, and is also interesting in the highest degree for its fossils, which are numerous, varied, and in excellent condition. The strata of the series have been arranged as Upper Oolite, Middle Oolite, and Lower Oolite. The Oolite forms the upper division of the Jurassic System, so-called because the range of the Jura Mountains is almost entirely composed of such limestones.

Opal

Opal is an amorphous, or non-crystalline form of silica and a relative hardness of 6. It is always a secondary mineral, found lining and filling cavities in igneous and sedimentary rocks where it has been deposited by hot waters. Opals are found in many parts of Europe, especially in Hungary, in the East Indies, etc. The ordinary varieties are common but the precious varieties are quite valuable. There are many varieties or species, the chief of which are: (a) precious or noble opal, which exhibits brilliant and changeable reflections of green, blue, yellow, and red; (b) fire opal, which simply affords a red reflection; (c) common opal, whose colours are white, green, yellow, and red, but without the play of colours; (d) semi-opal, the varieties of which are more opaque than common opal; (e) hydrophane, which assumes a transparency only when thrown into water; (f) hyalite, which occurs in small globular and botryoidal forms, with a vitreous lustre; (g) menilite, which occurs in irregular or reniform masses, and is opaque or slightly translucent. Opal is also a petrifying agent which may permeate fossil wood, corals, shells etc, preserving even the minute structure of the organic remains. It was discovered in 1964 that opal is made up of close-packed silica spheres, and in 1974 Pierre Gilson produced the first synthetic opal.

Ophite

Ophite or ophiolite is a green porphyry or serpentine, a metamorphic rock of a dusky-green colour of different shades, sprinkled with spots of a lighter green. It is a hydrous silicate of magnesia with alumina and iron.

Ordovician

The Ordovician period was the fourth geological period, 375,000,000 years ago.

Ore

Ore is any compound of a heavy metal which occurs naturally. The commonest are metallic sulphides, for example galena and iron pyrites. Oxides and carbonates are also frequently met with. If the metal occurs in the free state in nature it is said to be native. Metals are commonly obtained from their ores by smelting, the ores having been previously oxidized by roasting. Ores are commonly found on veins or lodes.

Orpiment

Orpiment is a secondary mineral with the formulae As_2S_3 and a relative hardness of 2. It is a rare mineral usually associated with realgar and occurs in laminated or lamellar masses, in concretions, and more rarely in minute crystals. It is also manufactured artificially. Orpiment is an important source of industrial arsenic and is used in dyeing. Orpiment can be distinguished from sulphur by its perfect cleavage, and strong smell of garlic (arsenic) when heated.

Orthite

Orthite is a silicate of aluminium containing the rare metals cerium, lanthanum, didymium, and yttrium, occurring in granite and other rocks in Sweden, Greenland, the Ural, etc.

Orthorhombic

Orthorhombic refers to a rectangular crystal with three axes of different lengths and all at right angles to each other. A closed book is a basic example.

Outcrop

In geology, an outcrop is the exposure of an inclined stratum at the surface of the ground.

Oxford Clay

Oxford clay or the Oxfordian beds, are the lowest subdivision of the Upper Jurassic rocks. Their outcrop stretches from Weymouth, past Swindon, Oxford and Bedford, into North Yorkshire. In its typical development it is 150 meters or more in thickness, blue or grey in colour, and contains thin layers of limestone and many septarian nodules.

Oxide

Oxide refers to a group of minerals where oxygen joined with a metal is a major constituent.

Oxidized Zone

Oxidized zone refers to the portion of an ore body that has been altered by downward percolating groundwater and which contains dissolved oxygen and carbon dioxide.

Ozocerite

Ozocerite is a wax-like mineral resin. It is sometimes called native paraffin, or mineral wax.

Ozokerite

Ozokerite is a fossil resin of a pleasantly aromatic odour, existing in the bituminous sand-stones of the coal measures, and occurring chiefly in Galicia, in Austria. Small quantities of it were found at Uphall and at Urpeth Colliery, Newcastle-on-Tyne, and various other places. It contains carbon and hydrogen in the proportion of 86 per cent of the former to 14 per cent of the latter. When purified it forms a hard paraffin, from which excellent candles were manufactured. It was also formerly used to some extent as an adulterant of bees-wax.

Palaeocene

The Palaeocene was the thirteenth geological period, 80,000,000 years ago.

Palagonite

Palagonite is a fine dark red rock, usually found in thin layers between ancient flows of basaltic lava. It is really a weathered and oxidised basic tuff or ash-bed, composed of shreds of black volcanic glass, pumice and rock fragments.

Pearly

Pearly refers to a lustre with the iridescent look of a pearl. Most commonly seen on surfaces parallel to cleavage planes.

Pebble

A pebble is a class of rock between 2 and 64 mm in diameter.

Pectolite

Pectolite or Osmelite is a white or greyish-white brittle mineral, a hydrous silicate of sodium and calcium with the formulae $NaCa_2Si_3O_8(OH)$ and a relative hardness of 5. It is formed from hydrothermal solutions filling cavities in basalts. Pectolite was confirmed as a distinct species in 1828 and is associated with zeolites, prehnite and calcite.

Pediment

In physical geography, a pediment is a broad, gently sloping rock surface that extends outwards from the abrupt foot of a mountain. Pediments are found in arid and semi-arid regions and are usually covered with a layer of alluvium.

Pediocratic

In geology, the term pediocratic refers to a period during which erosion predominates over crustal upheaval, leading to an overall lessening of relief.

Pedion

In crystallography, a pedion is a crystal form consisting of a single face, without any symmetrically equivalent face.

Pegmatite

Pegmatite is a variety of extremely coarse-grained igneous rock chemically similar to and closely associated with granite. The mineral constituents ofpegmatite are largely those typical of acidic intrusive rocks: orthoclase feldspar, quartz, and mica. Individual mineral crystals may be up to several meters in length. The growth of crystals this large indicates that the parent magma, from which the rock solidified, was able to cool very slowly, after injection into fissures extending outward from a central magma chamber.

Penninite

Penninite is a member of the chlorite group of minerals.

Pentlandite

Pentlandite is the most important industrial source of nickel and a relative hardness of 4. Pentlandite was confirmed as a distinct species in 1856 and named after Joseph Pentland, the discoverer of it. Pentlandite forms in basic igneous rocks such as diabase and gabbro through magmatic segregation.

Peridote

Peridote is a yellow variety of tourmaline found in Sri Lanka.

Peridotite

Peridotite is an intrusive igneous rock thought to be one of the major components of the Earth's mantle. Peridotite is generally dark coloured and formed from dense, heavy minerals, of which more than 40 percent are olivine.

Perlite

A perlite is an obsidian, or other vitreous rock with a concentric structure and which is expansible by heating.

Perlitic Structure

A perlitic structure in volcanic rocks consists of small concentric cracks, along which the rock readily breaks down, yielding rounded pearl-like fragments. The phenomena is found in some pitchstones and obsidians.

Permian

The Permian was the eighth geological period, 205,000,000 years ago. It marked the evolution of the reptiles.

Perthite

Perthite is a variety of orthoclase feldspar or microcline in which small veinlets of plagioclase feldspar (albite) are present. It is a common mineral in rocks such as granite and syenite.

Petalite

Petalite has the formulae $LiAlSi_4O_{10}$ and a relative hardness of 7. It is an ore of lithium. Associated with spodumene, lepidolite, tourmaline.

Petroleum

Petroleum is a mineral oil from which petrol is derived.

Pharmacosiderite

Pharmacosiderite is a secondary mineral formed through the oxidation of pre-existent arsenic-bearing minerals such as arsenopyrite. It is also commonly found associated with goethite, pyrite, realgar and scorodite. Pharmacosiderite was confirmed as a distinct species in 1813.

Phlogopite

Phlogopite has the formulae $K(Mg,Fe)_3(AlSi_3)O_{10}(F,OH)_2$ and a relative hardness of 3. It occurs as a result of the metamorphism of crystalline magnesium limestones or dolomitic marbles. Also found in serpentine. Rarely found in igneous rocks.

Phonolite

Phonolite or clinkstone is a volcanic rock, consisting essentially of nepheline and sanidine feldspar. It usually also contains some form of augite, hornblende or biotite.

Phosphates

Phosphates refers to a group of minerals where phosphate (PO_4) is an important constituent.

Phosphorite

Phosphorite is an impure massive form of apatite, mined for the calcium phosphate it contains, and forming one of the chief sources of phosphorus and its compounds, and of artificial manures.

Phyllite

Phyllite is a clay rock, resembling slate, but containing a large percentage of fine, silky, pale- coloured mica. It splits readily, and has a shimmering lustre on the surface.

Picrite

Picrite is a crystalline igneous rock which consists essentially of olivine and augite, but usually also contains iron oxides, a little feldspar, biotite, hornblende and apatite.

Piemontite

Piemontite (piedmontite) is a rare reddish-brown coloured mineral first discovered in New Zealand and confirmed as a distinct species in 1853. It is a hydrous silicate of calcium, aluminium and ferric iron and usually contains between five and twenty percent manganese oxide. Piemontite is found in metasomatic manganese ore deposits in andesite, rhyolite and low-grade schist. Rarely it also occurs in pegmatite. Piemontite has a relative hardness of 6.

Pipe

Pipe refers to a cylindrical, vertical mass of igneous rock.

Placer

Placer refers to a concentrated deposit of mineral particles that have weathered out of rock. Usually deposited by stream action.

Plagioclase

Plagioclase has the formulae $Na(AlSi_3O_8)$ and a relative hardness of 6. It is the plagioclase feldspar group includes: albite, oligoclase, andesine, labradorite, bytownite, anorthite.

Platinum

Platinum is a rare, silvery coloured, heavy metallic element with the symbol Pt. It is relatively inert and has a high melting point. Platinum is used in jewellery and as a catalyst.

Plattnerite

Plattnerite is a black and heavy secondary mineral of the rutile group formed through the alteration of primary lead-bearing minerals and the oxidation of lead itself. Plattnerite was confirmed as a distinct species in 1845 and named after K F Plattner, professor of metallurgy and assaying at the University of Freiburg,Germany. Plattnerite is a lead oxide and has a brilliant metallic lustre and tetragonal crystals. Plattnerite is closely related and is dimorphous to scrutinyite.

Playa

Playa refers to a shallow basin or plain in a desert where water collects after a rain and then evaporates.

Pleistocene

The Pleistocene (ice age) was the seventeenth geological period, 500,000 years ago. It marked the evolution of tool-making man.

Pliocene

The Pliocene was the sixteenth geological period, 8,000,000 years ago. It marked the modern formation of the continents.

Plutonic Rock

Plutonic rock is igneous rock that has solidified at depth in the Earth's crust. Having cooled slowly, plutonic rocks are coarse-grained: granites and gabbros are typical plutonic rocks. A body of plutonic rock can be called a pluton, but this word has also been used in a narrower sense for a granite mass occurring at a relatively high level in the crust and more or less circular in plan.

Porter

Porter (black and gold) is a type of marble found in the Apennines. Porter has a black background clouded with grey and is strongly veined with chain-like formations in yellow - whence it's alternative name of 'black and gold'.

Potash

Potash is the name given to any mineral containing potassium.

Potassium Argon Dating

Potassium argon dating is a technique used in geology for estimating the age of a mineral or rock, based upon the rate of decay of radioactive potassium into argon.

Potstone

Potstone (*Lapis ollaris*) is a species of talc containing an admixture of chlorite. It is green in colour and varies in shade, greasy and soft but hardens on exposure to the air. It's English name comes from its capability to be turned into vases and pots by turning.

Pre-Cambrian

The Pre-Cambrian period was the second geological period. It was the time when the first animal life appeared, about 1,125,000,000 years ago.

Prehnite

Prehnite is a mineral named after Colonel Hendrick Von Prehn who first discovered it at the Cape of Good Hope in South Africa. It was confirmed as a distinct species of mineral in 1788. Prehnite is a hydrous silicate of calcium and aluminium with the formulae $Ca_2Al_2Si_3O_{10}(OH)_2$ and a relative hardness of 6 to 7. Prehnite occurs as a crusty lining in cavities in basalt and related rocks and is associated with zeolites, datolite, pectolite, and calcite. Prehnite has a characteristic green colour. Resembles hemimorphite but is of a lower specific gravity and fuses easily.

Propylite

In geology, a propylite is an altered andesite or similar rock containing calcite, chlorite, etc., produced by the action of hot water.

Proustite

Proustite is an ore of silver with a characteristic ruby-red colour, vermilion streak and a brilliant lustre. It has a relative hardness of 3.

Pseudomalachite

Pseudomalachite is a rare green coloured mineral named for its superficial resemblance to malachite. Both minerals contain copper, but pseudomalachite is a hydrous hydrated phosphate rather than a hydrous carbonate. Pseudomalachite was confirmed as a distinct species of mineral in 1813 after being discovered at Virneberg in Germany. Pseudomalachite has a relative hardness of 4.5 to 5.

Pseudomorph

Pseudomorph refers to a mineral that has taken the outward crystal form of a different mineral.

Pyrite

Pyrite is one of the most common minerals. It is a shiny yellow mineral resembling gold, whence it's name 'fool's gold' formed from cooling magma and is found as an igneous segregation and also in metamorphic rocks and as vein deposits. Often it is found in sedimentary rocks being both primary and secondary in origin. Pyrite is often mined for the gold or silver associated with it. Pyrite has the formulae FeS2 and a relative hardness of 7. During the Second World War pyrite was mined in the USA for its sulphur which was used to produce sulphuric acid.

Pyroclast

In mineralogy, a pyroclast is a fragment of volcanic rock scattered by an explosive eruption.

Pyroxenes

Pyroxenes refers to a group of closely related and dark coloured rock forming minerals. Examples are aegirine, diopside, enstatite, hedenbergite, jadeite and spodumene. Pyroxenes are common rock-forming minerals which are found in most igneous and many metamorphic rocks. Their presence in a rock indicates that it crystallised at a high temperature.

Pyroxenite

Pyroxenite is a medium to coarse-grained intrusive igneous plutonic rock made up almost completely of pyroxenes. Containing less than 45 per cent silica, pyroxenite contains large quantities of iron- and magnesium-bearing minerals.

Pyrrhotite

Pyrrhotite is a magnetic common mineral often found in igneous rocks, and confirmed as a distinct species of mineral in 1835. It is an iron sulphide with the formulae FeS and a relative hardness of 5. It is mined for the nickel minerals associated with it.

Quartz

Quartz is a very widespread mineral which occurs as an important constituent of igneous rocks which have an excess of silica. Very resistant to mechanical and chemical attack. Massive, fine grained types are called jasper, chert, flint, agate. Coarse crystalline types called amethyst, rose quartz etc. Quartz has the formulae SiO_2 and a relative hardness of 7. During the Second World War attempts were made to synthesise quartz, and in 1947 the first optical grade crystals large enough for scientific use were grown by the hydrothermal method.

Quartz Porphyry

Quartz porphyry is a widely distributed intrusive igneous rock formed from magma in small (relatively speaking, an area of 5000 square kilometres is 'small' in geological terms) intrusive structures, such as dikes and sills, which were usually formed at no great depth. It is generally light coloured, usually grey, pink, red or violet in colour, and contains more than 65 per cent silica and more than ten percent quartz, as feldspar, mica and quartz.

Quartzite

Quartzite is a rock that occurs in two forms: sedimentary quartzite, also known as orthoquartzite; and metaquartzite which occurs as a result of sedimentary quartzite being changed by heat. Quartzite forms from pure sands on the beds of shallow seas. Quartzite is a tough rock comprised of tightly compressed grains, and is used to make flooring and facings in the construction industry.

Realgar

Realgar is a sulphide of arsenic found in veins of lead, silver, and gold ores associated with arsenic minerals and stibnite and also occurs as a deposit from hot springs. When mixed with saltpetre and burned it gives a bright white light and was formerly used in fireworks to provide colour. It has the formulae AsS and a relative hardness of 2.

Red Crag

Red Crag are deposits of fossil remains on the coast of Essex and Suffolk in England used in the manufacture of manure.

Red Ochre

Red Ochre is a soft, earthy variety of haematite mineral.

Replacement

Replacement refers to the process by which one mineral is replaced by another and the original physical form is often retained.

Resinous

Resinous refers to a lustre with the appearance of resin.

Rhaetic Beds

The Rhaetic beds are a group of strata which occur at the top of the Triassic and at the base of the Jurassic system. They are also known as the White Lias, or the Avicula contorta beds, from one of their best known and commonest fossils. In England they may be traced from the coast of Devon past Bristol to Redcar on the Yorkshire coast, forming a narrow strip on the west side of the Lias shales. They contain comparatively few fossils; but their fauna, though small, is very characteristic, and includes several forms that do not occur elsewhere.

Rhodochrosite

Rhodochrosite is a carbonate of manganese and an important manganese ore. It occurs in veins of copper, lead and silver that have been altered by hydrothermals, and as a secondary deposit in sedimentary layers of manganese oxide. Rhodochrosite is mainly associated with calcite, but may also occur with quartz. In South America it is used as an ornamental stone. It has a relative hardness of 3.5 to 4.

Rhodonite

Rhodonite is an often pink or red coloured mineral sometimes polished for use as an ornamental stone. Rhodonite is formed mainly in manganese-bearing metamorphic rocks, especially marble and skarn, which were previously limestone. It is also found in sediments that have been altered by metasomatism. It has the formulae (Mn,Fe,Mg)SiO_3 and a relative hardness of 7.

Rhyolite

Rhyolite is a fine grained extrusive igneous rock similar to granite. Rhyolite forms from particularly viscous molten lava and is grey, brown or reddish in colour, darkening with age. Rhyolite is essentially composed of feldspar, mica and quartz with numerous accessory minerals present.

Riebeckite

Riebeckite is an accessory mineral of the amphibole group found in granite and syenite and other igneous rocks. Riebeckite is a hydrous silicate of sodium, ferrous iron, magnesium and ferric iron named after the German Emil Riebeck who first discovered it.

Rock Cork

Rock cork is a species of asbestos, resembling cork. It is soft, easily cut and very light.

Rocks and Minerals

A mineral is a homogeneous inorganic natural substance of a definite chemical composition, it is usually a solid, but occasionally a liquid. A few are composed of a single element, but most are formed by the combination of several elements. A product which has developed by a natural process is a natural substance. A mineral is homogeneous physically and chemically when it consists of elements of the same kind throughout. In this a mineral differs from rock, which is also an inorganic natural substance, but one composed of different minerals, and which, in contrast to them, forms large parts of the earth's crust. There are some rocks (for instance sandstone or limestone), which consist almost entirely of one mineral (quartz or calcite for example). We do not classify them as minerals, for they do not have to be homogeneous, as do minerals. The science which studies minerals is called mineralogy. It not only observes and defines the various characteristics of minerals, but also explains their origin and development. Petrography concentrates on studying rocks. These sciences, together with geology, which studies the origin, composition and evolution of the earth, are not just theoretical sciences. Their findings affect practical life with ever increasing force. The vast and vital role that minerals and rocks play in the modern world is taken for granted. They form the basic raw materials of industry, whether they are metallic minerals extracted from ores, minerals for the chemical, glass or ceramic industry, or others.

Rose Diamond

In jewellery, a rose diamond is a diamond that is quite flat underneath, with its upper part cut in diverse little faces, usually triangles, the uppermost of which terminates in a point.

Rubasse

Rubasse is a beautiful spangled red variety of quartz, coloured by the presence of minute blood- coloured scales of iron oxide. The natural stone is very rare, but may be found in Brazil.

Ruby

Ruby is the red transparent form of corundum containing impurities of chromium and iron. It is a precious stone. Artificial rubies have been manufactured since 1902 from ammonia alum and chrome alum. And in 1960 and artificial ruby was used in the first working laser.

Rutile

Rutile is an accessory mineral found in igneous rock and in metamorphic gneiss and schist. It is an ore of titanium, and is a high-temperature form of titanium dioxide. It has the formulae TiO_2 and a relative hardness of 6.5. It exhibits a strong lustre and is often found in twinned crystals. Rutile is found in granite, granite pegmatites, gneiss, mica schist, metamorphic limestone and dolomite. It is present as an accessory mineral in the rocks or in quartz veins. The name rutile is derived from the Latin word 'rutilus' meaning red.

Salina Beds

The Salina Beds belong to the Upper Silurian of North America, and are the lower subdivision of the Onondaga group. They consist mainly of red marls with limestone, dolomite, and extensive beds of rock-salt, especially in the state of New York. As a whole, these rocks are unfossiliferous, and seem to have been laid down in shallow waters exposed to desiccation. They extend northwards into Ontario, crossing the Niagra river a little above Niagra Falls.

Sand

Sand is small particles (between 0.0625 and 2 mm in diameter) of mineral matter, usually quartz. The purest quartz sands are white in colour and used for making glass. Other sands may be various colours dependant upon the minerals they contain.

Sandstone

Sandstone is a common sedimentary rock made up of tiny grains of quartz held together by a type of cement. Sandstone is produced by sand which has been deposited and cement precipitated between the grains by groundwater percolating through it. The most common cements are calcium carbonate, silica and iron oxide. Sandstone is an important building material in the USA where many of the brownstone houses in the north-east cities are made of sandstone.

Sanidine

Sanidine is a clear, glassy, often cracked variety of orthoclase feldspar, which is confined to modern volcanic rocks, such as trachyte, rhyolite and phonolite. It is found in the Rhine, Italy and Hungary.

Sapphire

Sapphire is a coloured form of corundum, often blue and transparent. The colouration is caused by the presence of iron oxide and titanium oxide. Artificial sapphires are made by melting powdered corundum at high temperature with iron and titanium.

Sard

Sard is a brownish red variety of Chalcedony.

Sassolite

Sassolite (Boric Acid) is a vitreous, white mineral found in volcanic areas and in hot springs. It is named after Sasso in Italy where it was first discovered.

Saussurite

Saussurite is a waxy-looking, white or pale green substance which results from the alteration of feldspar in rocks such as gabbro. It is a very fine admixture of garnet, zoisite, albite and other minerals.

Scapolite

Scapolite a relative hardness of 6. It shows fluorescence. Occurs in the crystalline schists, gneisses, and often is probably derived from the alteration of plagioclase feldspars. Also occurs in crystalline limestones formed by metamorphic contact or igneous intrusion. Associated with diopside, amphibole, garnet, apatite, and zircon.

Scheelite

Scheelite is a mineral occurring in hydrothermal veins and in placer deposits, in contact metamorphic rocks and in pegmatite. It an important ore of tungsten.

Scheelite was confirmed as a distinct species of mineral in 1821 and named in honour of the Swedish chemist Carl Scheele. Scheelite is associated with cassiterite, topaz, fluorite, apatite, molybdenite, and wolframite. Sometimes scheelite is found with gold. It has the formulae $CaWO_4$ and a relative hardness of 5.

Schist

Schist is a medium to coarse-grained metamorphic rock formed mainly from finely layered sheets of platy or elongated minerals such as mica. These sheets are interleaved with feldspar and quartz.

Schorl

Schorl is a rock comprised of quartz and black tourmaline (schorl). It is of igneous origin and occurs associated with granite and crystalline schists. It has a granular texture and is usually a grey colour. It is a frequent associate of tin-bearing veins in all parts of the world.

Schwazite

Schwazite is a variety of Tetrahedrite containing up to 17 per cent mercury.

Scolecite

Scolecite is a mineral found in lavas and contact metamorphic deposits. It is a hydrated silicate of calcium and aluminium and has the formulae $Ca(Al_2Si_3)O_{10}3H_2O$ and a relative hardness of 6. Scolecite was confirmed as a distinct species of mineral in 1813.

Scoria

A scoria is a cavity in igneous rock, caused by an escape of gas or steam. When such a cavity contains a mineral it is known as a amygdule.

Scorodite

Scorodite is a mineral of the varicite group occurring in the oxidized portions of metallic veins as an alteration of arsenic containing minerals. It has the formulae $FeAsO_42H_2O$ and a relative hardness of 4. Scorodite was confirmed as a distinct species of mineral in 1818.

Secondary Minerals

Secondary minerals refers to minerals formed by the alteration of pre-existing minerals.

Sedimentary Rock

Sedimentary rock is a deposit that has been laid down by water, wind, ice, or gravity. The material of which sedimentary rocks are composed has usually been transported from its source. They have accumulated at or near the Earth' s surface at normal temperatures and pressures, and are thus distinguished from igneous rocks, which have originated as magma in a molten or semi- molten state, and from metamorphic rocks, which have been produced by the effects of heat or pressure. Many sedimentary rocks were deposited in water in which the transported material was in a state of suspension or solution. After deposition they have been consolidated. Some, such as loess, were deposited by the wind; some, such as boulder clay, are the result of glacial action; some have simply accumulated in place. Sediments can be divided into three groups: clastic rocks, chemical precipitates, and organic sediments. Of these, clastic sediments are the most abundant. They are made up of fragments of pre-existing rocks. Clays, sands, and gravels are in this category.

Selenite

Selenite is a clear, transparent variety of gypsum often found in large twinned crystals shaped like arrow-heads or rhombus. The crystals have a perfect cleavage like mica, but are less flexible and somewhat softer. Selenite has a relative hardness of 2 and the formulae $CaSo_{42}H_2O$.

Serpentine

Serpentine is a common mineral group with a relative hardness of 4. It is usually an alteration product of some magnesium silicate, especially olivine, pyroxene, and amphibole. It is frequently associated with magnesite, chromite, and magnetite.

Shale

Shale is a fine black sedimentary rock that splits easily into thin plates along well-defined planes parallel to the original stratification. Shale is composed of a complex mixture of microscopic clay minerals plus mica and quartz. Shale is often rich in fossils and dinosaur tracks. Shale is formed in quiet water environments, such as lake bottoms, river floodplains, tidal flats, marshes and the floors of deep oceans, where the finest particles carried by the water have settled. These clays gradually petrify forming shale.

Siderite

Siderite is native iron carbonate, an ore of iron that becomes magnetic when heated. It is frequently found as an impure admixture with clay materials and also in concretions with concentric layers. It has a relative hardness of 5.

Silica

Silica refers to silicon dioxide (SiO_2). A very common mineral that is found in many forms including quartz, opal, chert.

Silicates

Silicates refers to a group of minerals composed chiefly of SiO_4. Ex: quartz, orthoclase.

Silky

Silky refers to a silk-like lustre on a mineral. Results from a fine, fibrous and parallel surface.

Sill

Sill is a geological term for a sheet of igneous rock intruded into overlaying beds and lying nearly horizontal over a large area.

Sillimanite

Sillimanite (fibrolite or bucholzite) is a fibrous silicate, somewhat rare mineral found as a constituent of gneiss and schist in metamorphic rocks, and often occurs with corundum. It has the formulae Al_2SiO_5 and a relative hardness of 7. Sillimanite was first confirmed as a distinct species of mineral in 1824 and was named in honour of Benjamin Silliman, professor of chemistry at Yale University. Industrially, sillimanite is used in spark-plugs for automobiles.

Silt

Silt is a class of rock between 0.004 and 0.0625 mm in diameter.

Silurian

The Silurian was the fifth geological period, 335,000,000 years ago. This period marked the appearance of the first land plants.

Silver

Silver is a white, ductile metallic element that is sonorous, very malleable, capable of a high degree of polish, and chiefly univalent with it's compounds and has the highest thermal and electrical conductivity of any substance. It has the symbol Ag.

Skarn

Skarn is a calcite rock formed from pre-existent limestone and dolostone when molten magma escapes from the Earth's mantle and becomes injected into the upper crust where it forms igneous intrusions altering the existing rocks by its heat. Skarn is often rich in minerals.

Skutterudite

Skutterudite is an important ore of cobalt and nickel usually formed in hydrothermal veins. It is associated with native silver, bismuth, calcite and arsenopyrite. Skutterudite is an arsenide of cobalt, iron and nickel and has a relative hardness of 6. Skutterudite was confirmed as a distinct species of mineral in 1845.

Slate

Slate is a greyish metamorphic rock. Slate is commonly found where shale has been subjected to low-grade regional metamorphism of low pressure and very little heat during episodes of mountain building. It is very fine grained and splits well into thin sheets. It is composed of mica, quartz and other minerals and is used for roofing tiles and sheets.

Slickensides

In geology, slickensides are the wall-faces or polished surfaces found along the joints and fault planes of rocks. They are usually marked with fine striations of the sides of the fissure over one another.

Slime

Slime is the name applied to that portion of crushed ore which, owing to its impalpable and minutely divided state and the presence of colloidal substances, settles very slowly from water, forming an impervious mass.

Smithsonite

Smithsonite is a secondary mineral formed in parts of zinc and copper deposits that have been oxidised by the action of air and water. It is one of the calamines and was confirmed as a distinct species of mineral in 1832 and named after James Macie Smithson the British chemist and mineralogist. Smithsonite is grey, white, light green or brown in colour with a glassy lustre. It has the formulae $ZnCO_3$ and a relative hardness of 6. It is an ore of zinc usually found with zinc deposits in limestone beds.

Snake Stone

Snake stone (also known as Water of Ayr Stone or Scotch Stone) is a variety of whetstone. It is a light grey carboniferous shale, used for polishing marble and copper plates as well as sharpening tools. Snake stone is found at Bridge of Stair on the Ayr in Scotland.

Soapstone

Soapstone is a popular name for steatite.

Sodalite

Sodalite is a usually attractive blue colour mineral found in limestones and used for polished slabs and for carved ornaments. It is a chlorosilicate of sodium and aluminium with the formulae Na4Al3(SiO4)3Cl and a relative hardness of 6. Sodalite was confirmed as a distinct species of mineral in 1811 and gets its name from its chief metal - sodium.

Soil creep

Soil creep is the name given to slow movements of the particles of soil along sloping surfaces. Gradual loosening of upper layers of soil is caused by the action of frost, rain and the weather in general, which assisted by gravity aids in bringing loosened material from higher to lower levels. Soil creep in the Culebra cut interfered with the continuous use of the Panama Canal when it was first built.

Solenhofen Stone

In geology, Solenhofen Stone is a name given to a fine-grained smooth-textured limestone, derived mainly from Solenhofen in Bavaria. The stone is found in beds of 15 meters or more thick, and is a deposit of the Upper Jurassic, and is particularly rich in fossils, notably those of pterodactyls and the archaeopteryx. Solenhofen Stone was widely used in lithography.

Sorbite

Sorbite is a compound of iron and carbon intermediate between austenite and pearlite, virtually an extremely minute form of the latter. Sorbite is named after the English geologist Henry Sorby.

Specific Gravity

Specific gravity refers to the relative density of a mineral. It is the ratio of:; Weight in Air/ (Weight in Air - Weight in Water). ;This measurement is an easily accomplished procedure using a simple balance or spring scale.

Sperrylite

Sperrylite is a rare natural compound of platinum and arsenic found in the nickel mines of Sudbury in Canada. It has the formulae PtAs2 and a relative hardness of 7.

Spessarite

Spessarite is a variety of garnet of a dark red colour found in the Ural Mountains and in Virginia, USA.

Sphalerite

Sphalerite or blende, is the main ore of zinc. Sphalerite is associated with galena, pyrite, marcasite, chalcopyrite, calcite. Sphalerite is formed mainly in hydrothermal veins and as a replacement deposit in limestones and in veins in igneous rocks. It has the formulae (Zn,Fe)S and a relative hardness of 4.

Sphene

Sphene (titanite, calcium silicate or titanate) is a variety of titanite and a source of titanium. A rather common accessory mineral in igneous rocks, it is a light yellow or green colour and sometimes a dark brownish green colour. It is often found as crystals, particularly in the Tirol, and is commonly associated with chlorite. It has the formulae $CaTiO(SiO_4)$ and a relative hardness of 6.

Spinel

Spinel is a common metamorphic mineral occurring imbedded in crystalline limestone, gneisses, and serpentine. Spinel occurs as a common accessory mineral in many dark igneous rocks. When transparent and finely coloured it is often confused for ruby and is used as a gem stone. An oxide of magnesium and aluminium it has the formulae $MgAl_2O_4$ and a relative hardness of 8.

Spodumene

Spodumene is a mineral confirmed as a distinct species in 1800 it is the main source of the metal lithium. Spodumene is found occasionally as very large crystals in pegmatic dikes. A silicate of lithium and aluminium, it has the formulae LiAlSi2O6 and a relative hardness of 7.

Stack

In geography, a stack describes an isolated column close to the sea shore. In the erosion of cliff faces fragments are left separated from the rock face, and may be honeycombed with caves, and will ultimately disappear. The Needles off the west extremity of the Isle of Wight are a conspicuous example of stacks.

Stalactite

A stalactite is a mass of calcareous matter hanging in caves, formed by the filtration of water containing calcium bicarbonate in solution through holes or pores in the cave roof. the evaporation of the water and carbonic acid gas leaves behind it a deposit of limestone which continues to increase in size so long as the water drops.

Stalagmite

A stalagmite is a column of calcium carbonate rising from the floor of a cave, formed by water containing the mineral falling from the roof or from a stalactite.

Stannite

Stannite or tin pyrites is a sulphostannate of copper and iron found in tin deposits in Cornwall.

Staurolite

Staurolite is an brown, vitreous accessory mineral found in crystalline schists, slates, and sometimes gneisses and often associated with garnet, kyanite, and tourmaline. It has relative hardness of 8.

Staurolite was confirmed as a distinct species of mineral in 1792 and often forms in distinctive cruciform shapes from which it is named.

Steatite

Steatite (soapstone) is a variety of talc, hydrated magnesic silicate, usually devoid of a distinct crystalline structure. It is white, blue-grey, green or brown in colour, soft and easy to cut and heat-resisting. Since the earliest of times steatite has been used for ornaments and carvings. In a powdered form it is an ingredient in soaps, paints, lubricants and French chalk.

Stellerite

Stellerite is a mineral of the zeolite group closely resembling stilbite. Stellerite was confirmed as a distinct species of mineral in 1909 after years of being confused with stilbite, and was named after Georg Steller the German explorer and naturalist.

Stephanite

Stephanite is a minor ore of silver. One of the last minerals to form in silver veins. It has the formulae Ag5SbS4 and a relative hardness of 3.

Stibiconite

Stibiconite is a secondary mineral formed through the oxidation of stibnite. It commonly occurs with stibnite and is sometimes pseudomorphous with it.

Stibnite

Stibnite (also known as antimonite, antimony glance and black antimony) is the major ore of antimony. It is Deposited by alkaline waters, usually in association with quartz. Stibnite is found in quartz veins or beds in granite and gneiss and may occur as a replacement in limestones and shales.

Stilbite

Stilbite is a secondary mineral formed by the action of hydrothermals on pre-existent minerals. It is found in cavities in basalts and related rocks in Scotland, Ireland and the Faroe Islands. It has a relative hardness of 4. Stilbite is a member of the zeolite group and was confirmed as a distinct species of mineral in 1797. Stilbite is a hydrated silicate of sodium, calcium and aluminium and is white to red in colour and exhibits a glassy or pearly lustre.

Stonebrash

Stonebrash is a soil made of small stones or broken rock. The term originated for the subsoil of north-western Wiltshire which consisted of a reddish calcaerous loam mingled with flat stones.

Stonesfield Slate

Stonesfield Slate is a limestone of the Jurassic period. Stonesfield Slate is not a slate, but is so called from being thin bedded, and can be used as a replacement for slate in many instances.

Stratification

In geology, stratification refers to the natural formation of layers or 'beds' of rock by the deposit of sediment. Many stratified rocks were formed by sediment washed down rivers into the sea where they formed a deposit on the sea bed and were subsequently covered by other deposits and compressed into rock. A common example being sandstone.

Stratum

Stratum is a geological term describing a mass of sedimentary rock (a strata) of great horizontal extent, which was deposited more or less continuously on the bottom of former seas or lakes, or sometimes on the surface of flat plains or deserts. Stratified rocks are mostly sandstones, shales and limestone.

Streak

Streak refers to the colour of the powder produced when a mineral is rubbed over the surface of a piece of unglazed, white porcelain.

Striations

Striations refers to very small parallel grooves or narrow channels of the faces of a crystal.

Strontianite

Strontianite is a natural source of strontium, named after Strontian in Argyleshire where it was discivered in 1792 by Hope. Physically it is similar to cerussite and witherite. It is associated with barite, celestite, and calcite in veins in limestone. Occasionally it is found in igneous rocks and as a gangue mineral in sulphide veins. It has the formulae $SrCO_3$ and a relative hardness of 4.

Strunzite

Strunzite is a mineral confirmed as a distinct species in 1957, and previously thought to be laurite. Strunzite has the same chemical makeup of laurite and pseudolaurite but has a monoclinic crystal system. Strunzite is a hydrated phosphate of manganese and ferric iron and appears as coatings and tufts on the surface of granite pegmatite where it formed through the alteration of primary triphylite.

Sulphates

Sulphates refers to a group of minerals in which sulphate SO_4 is an important part.

Sulphides

Sulphides refers to a mineral group where sulphur is combined with one or more metals. The sulphides form an important class of minerals that includes the majority of the ore minerals. With them are classed the similar but rarer sulpharsenides, arsenides, and tellurides. Most of the sulphide minerals are opaque with distinctive colours and characteristically coloured streaks. Those that are non-opaque, such as cinnabar, realgar, and orpiment, have high refractive indices and transmit light only on thin edges.

Sulphur

Sulphur (brimstone) is a non-metallic element that occurs either free or in combination with sulphates and sulphides, is a constituent of proteins, exists in several allotropic forms including yellow orthorhombic crystals, resembles oxygen chemically but is less active and more acidic, and is used especially in the chemical and paper industries, in rubber vulcanisation, and in medicine for the treating of skin diseases. It has the symbol S. Sulphur has been known since ancient times, but was only discovered to be an element in 1809.

Sunstone

Sunstone is a form of translucent feldspar found in Siberia, Norway and North America, emitting a brilliant red metallic glitter from the background like hidden fire. Sunstone is a semi-precious gemstone and is usually cut with a flat or slightly convex surface.

Syenite

Syenite is a plutonic igneous rock consisting essentially of alkali-feldspar and one or more ferro-magnesian minerals, such as augite, hornblende, or mica. Syenite is often mistake for granite because of its colour, but has a lower silica content.

Sylvanite

Sylvanite is a rare ore of gold, silver, and tellurium of a steel grey, silver white, or brass yellow colour, formed in low temperature hydrothermal veins. It often occurs in implanted crystals resembling written characters, and hence is called graphic tellurium. It is associated with calaverite and other tellurides - usually in quartz gangue. It has a relative hardness of 2.

Sylvite

Sylvite (sylvine) is an industrial mineral, the naturally occurring potassium chloride, and as such the main mineral ore of potassium. Sylvite has a relative hardness of two and is used as a fertilizer. Sylvite is an evaporate mineral found in sedimentary evaporate deposits such as dried-out salt lakes and seas in deserts, and is associated with halite. Sylvite was confirmed as a distinct species of mineral in 1832 but was known at least during the 17th century.

Syncline

In geology, a syncline is the dipping of the strata inwards towards the axis of an earth fold, producing a trough-shaped or basin-like arrangement of the strata. Synclines form the troughs and anticlines the arches of a series of earth folds.

Table diamond

In jewellery, a table diamond is a thin diamond cut with a large, square, flat upper surface encompassed by four lesser faces.

Tachhydrite

Tachhydrite is a hydrous chloride of calcium and magnesium occurring in yellowish masses which rapidly deliquesce upon exposure. It is found in the salt mines at Stassfurt.

Tachylite

Tachylite (Tachylyte) is a natural glass, formed by the rapid cooling of molten basalt. It is a black or dark- brown and greasy looking substance. It is very brittle and occurs in basaltic obsidians in dikes, veins and intrusive masses.

Taconic

Taconic designates a rock pertaining to the series of rocks forming the

Taconic mountains in Western New England. They were once supposed to be older than the Cambrian, but later proved to belong to the Lower Silurian and Cambrian.

Talc

Talc has the formulae $Mg_3Si_4O_{10}(OH)_2$ and a relative hardness of 1. It is a mineral of secondary origin formed by the alteration of magnesium silicates. Usually found in metamorphic rocks where, in a non-crystalline form, it occurs as 'soapstone' - and can make up large rock masses. Used for laboratory table tops and for many industrial uses. It has a greasy feel.

Talus

A talus is a sloping heap of fragments of rock lying at the foot of a precipice.

Tantalite

Tantalite is a heavy mineral of an iron-black colour and sub-metallic lustre. It is essentially a tantalate of iron. Tantalite is formed in igneous rocks, especially granite pegmatite, and in placer deposits. It is associated with amblygonite, apatite, beryl, cassiterite, feldspar, quartz, spodumene and tourmaline. Tantalite is the main ore of tantalum.

Tectonic

Tectonic designates rock structures and external forms resulting from the deformation of the earth's crust, such as, tectonic arches or valleys.

Tellurite

Tellurite is a soft secondary mineral, an oxide of tellurium. It occurs sparingly in tufts of white or yellowish crystals formed through the oxidation of pre-existent tellurium compounds or of the element itself. Tellurite is most commonly associated with paratellurite with which it is dimorphous. Tellurite was confirmed as a distinct species of mineral in 1845 and is named after its main elemental component, tellurium. Tellurite has a relative hardness of 2.

Tenacity

Tenacity refers to the ability of a substance to resist being separated.

Tennantite

Tennantite, named after Smithson Tennant, an English chemist is a blackish lead-grey mineral, closely related to tetrahedrite. It is essentially a sulphide of arsenic and copper.

Tephrite

Tephrite is an igneous rock consisting essentially of plagioclase and either leucite or nephelite, or both.

Tephroite

Tephroite is an ash-grey coloured silicate of manganese.

Terrace

In geology, a terrace is a level plain, usually with a steep front, bordering a river, a lake, or sometimes the sea. Many rivers are bordered by a series of terraces at different levels, indicating the flood plains at successive periods in their history.

Terrane

In geology, the term terrane describes a group of rocks having a common age or origin. The term is nearly equivalent to formation, but is used somewhat less comprehensively.

Tertiary Period

In geology, the term tertiary period describes the first period of the age of mammals, or of the Cenozoic era.

Tetradymite

Tetradymite (telluric bismuth also known as bismuth telluride) is a telluride of bismuth. It is of a pale steel-grey colour and metallic lustre, and usually occurs in foliated masses mixed with sulphur and selenium.

Tetragonal

Tetragonal refers to a crystal with four rectangular(not square) sides and two square bases. A butter package is an example.

Tetrahedrite

Tetrahedrite (grey copper ore, fahlore, or panabase), so called because the crystals of the species are commonly tetrahedrons, is sulphide of antimony and copper, with small quantities of other metals. It is an ore of copper and silver. It is commonly found in hydrothermal veins formed at low to moderate temperatures and is usually associated with other silver, lead, and copper minerals. It has relative hardness of 5.

Thalassic

In geology, thalassic describes rocks pertaining to the sea. The term is sometimes applied to rocks formed from sediments deposited upon the sea bottom.

Thanet Beds

In geology, the Thanet beds is the name given to the lowest division of the Eogene in Britain. Also known as the Thanet Sands, they consist of yellow to greenish sands, well developed in the Isle of Thanet, and contain a large number of marine fossils.

The Age of Acrogens

In geology, 'the age of acrogens' is the age of coal plants, or the carboniferous era.

Thenardite

Thenardite, named after the French chemist, Louis-Jacques Thenard, is a white or brown coloured mineral with a vitreous lustre that dissolves easily in water and has a weak salty taste. It forms in sedimentary evaporate deposits in lakes and the playas of desert climates. Thenardite was confirmed as a distinct species of mineral in 1826 and is used in the glass and paper-making industries.

Thinolite

Thinolite is a calcareous tufa, in part crystalline, occurring on a large scale as a shore deposit about the Quaternary lake basins of Nevada.

Thomsenolite

Thomsenolite, named after Dr. J. Thomsen of Copenhagen, is a fluoride of aluminium, calcium, and sodium occurring with the cryolite of Greenland.

Thomsonite

Thomsonite (mesole, or comptonite), named by its discoverer, the English mineralogist Brook after his friend Thomas Thomson, of Glasgow, is a zeolitic mineral, occurring generally in masses of a radiated structure. Thomsonite is found fairly abundantly in cavities of phonolites and basalt rocks, more rarely in pegmatite veins. It is a hydrous silicate of alumina, lime, and soda.

Thorianite

Thorianite is a rare mineral found in Sri Lanka, and containing thorium, uranium, cerium and other rare earths. The percentage of throium oxide is as much as 70 in some specimens.

Thorite

Thorite, so called by Berzelius from the Scandinavian god Thor, is a mineral of a brown to black colour, or, as in the variety orangite, orange-yellow, and with a glassy lustre. It is essentially a silicate of thorium. Thorite is chiefly found in Norway and Sri Lanka.

Thrust

In geology, thrust is a name given to a type of fault or break in a rock formation. The term thrust is particularly applied to those faults the planes of which make a small angle with the horizontal. By thrust action masses of rock are pushed over underlying masses, sometimes for many kilometres. Examples of this type of action are apparent in Scandinavia, Scotland and the Appalachians.

Thrust Plane

In geology, a thrust plane is the surface along which dislocation has taken place in the case of a reversed fault.

Thulite

Thulite is a variety of zoisite used as an ornamental stone. It is a basic calcium and aluminium silicate, rose-red or pink in colour, and is found in Telemarken in Norway.

Thuringite

Thuringite, named after Thuringia where it is found is a mineral occurring as an aggregation of minute scales having an olive-green colour and pearly lustre. It is a hydrous silicate of alumina and iron.

Tigers-eye

Tigers-eye is a siliceous stone of a yellow colour and chatoyant lustre, obtained in South Africa and much used for ornament. It is an altered form of the mineral crocidolite.

Tile Ore

Tile ore is an earthy variety of cuprite.

Tilestone

A tilestone is a kind of laminated shale or sandstone belonging to some of the layers of the Upper Silurian.

Till

In geology, a till is a stiff unstratified deposit of clay mixed with sand, gravel and boulders without lamination, formed in a glacier valley by means of the waters derived from the melting glaciers. The term is sometimes applied to alluvium of an upper river terrace, when not laminated, and appearing as if formed in the same manner.

Titanium

Titanium is a silvery-grey coloured metal element (of the transition metals) resembling iron with the symbol Ti and atomic number 22. Titanium is the ninth most common element found in the earth's crust, occurring in accessory oxides and oxide minerals in igneous and metamorphic rocks and in beach sands (98 percent of rocks have been found to contain titanium). The principal sources of titanium are ilmenite (typically containing about 32 per cent titanium) and rutile (which in its pure state is titanium dioxide and contains a higher proportion of titanium than ilmenite.

Titanium was first discovered in 1791 by the British scientist William Gregor who named it menachanite after the town in Cornwall near where he found it, however titanium was independently discovered and named in 1795 by the German scientist Klaproth after he found it while investigating the mineral rutile. Klaproth named the metal titanium after the Titans of Greek mythology in view of the metal's strength (strangely, pure titanium is a soft and ductile material. It is only when combined in alloy form that it is so strong. Pure titanium has a tensile strength of 40,000 lbs per square inch, in alloy form this increases to 200,000 lbs per square inch). It took another 100 years before titanium could be separated, in 1910 by Hunter after which titanium became a laboratory curiosity until 1946 when the US Bureau of Mines succeeded in producing titanium on a pilot plant scale and its industrial importance was recognised. Titanium and has been produced commercially since 1948 and is renowned in industry for its lightness, strength and resistance to corrosion leading to its widespread use in alloys used in aircraft.

Titanium is produced by a chemical process in which titanium tetrachloride is reduced with magnesium or sodium. In the USA the Kroll process is used which consists of reacting four parts liquid with one part of molten magnesium in an iron reactor at about 850 degrees Celsius to form one part of titanium and four parts of magneisum chloride. In Britain titanium is usually produced with the sodium-reduction process which uses two parts of sodium to reduce four parts of titanium tetrachloride to produce one part of titanium and five parts of sodium chloride. In both reactions the atmosphere in the reactor is excluded of oxygen, hydrogen and nitrogen and replaced with argon gas to prevent contamination of the final product. Both reactions produce titanium in a sponge-like form (known as sponge titanium) with trapped crystals of the other substance which have to be removed by leaching with water and acid or being heated in a vacuum. The titanium sponges are then melted into ingots, but usually contain impurities of oxygen, carbon, nitrogen and iron.

Tonalite

Tonalite is a group of rocks consisting chiefly of quartz, orthoclase, plagioclase, hornblende and biotite. Apatite, magnetite, zircon, iron oxides etc are also often present in the rocks of the group. Tonalite resembles dark granite in its general appearance, and is so called from its occurrence at Monte Adamello, near Tonale in the eastern Alps. The rocks are also found in Scotland, Ireland and America.

Topaz

Topaz is a mineral aluminium fluosilicate occurring in rhombic prisms, generally yellowish and pellucid, also colourless, and of greenish, bluish, or brownish shades with a glassy lustre, fine varieties of the mineral are valued as gem stones. It sometimes occurs massive and opaque. Topaz is formed by fluorine-bearing vapours given off during the last stages of the solidification of igneous rocks and is frequently associated with tin stone, tourmaline, mica. beryl, etc. The deep orange yellow varieties of the mineral are most valued, and the finest stones have been found in the Ural Mountains and Brazil. Some of the Brazilian stones become pink on heating and are used in cheap jewellery. Topaz crystals are also found in Saxony, Scotland, Ireland, the south-west of England, Africa, India, Japan, North America, etc. Many so-called topazes are really only yellow varieties of quartz.

Topazolite

Topazolite is a topaz-yellow variety of garnet.

Torbanite

Torbanite (also known as boghead coal) is a dark brown variety of cannel coal found at Torbanehill in Scotland. Torbanite contains a large proportion of algae of the Palaeozoic age. Similar coal has been found in New South Wales. Torbanite contains over 60 per cent of volatile matter and is valuable for the extraction of burning and lubricating oils and paraffin.

Torbenite

Torbenite (chalcolite), named after the Swedish chemist Torbern Bergman, is a secondary mineral, a uranium ore, originating from the weathering of uranium. It contains up to 61 per cent uranium. It has a relative hardness of 2.5.

Tourmaline

Tourmaline is the most commonly found in granite pegmatites. It is usually black in colour, but lighter coloured gem varieties are also found. It occurs usually in three-sided or six-sided prisms terminated by rhombohedral or scalenohedral planes. Black tourmaline (schorl) is the most common variety, but there are also other varieties, as the blue (indicolite), red (rubellite), also green, brown, yellow (peridote) and white. The red and green varieties when transparent are valued as jewels. Crystals of tourmaline when heated exhibit electric polarity.

Trachyte

Trachyte is a group of igneous volcanic rocks, grey, yellow, brown, green and red in colour consisting chiefly of alkali feldspar, and often containing crystals of glassy feldspar, mica, hornblende, or augite. Trachyte occurs in lava, intrusive sheets, and dykes from the early Tertiary period. It is sometimes used as a paving material.

Trass

Trass is a white to grey coloured volcanic tufa, formed of decomposed trachytic cinders, sometimes used as a cement.

Travertine

Travertine is a white or light-coloured crystalline concretionary limestone, usually hard and semi crystalline, deposited from springs and rivers and used for building. Extensive deposits exist at Tivoli, near Rome.

Tremolite

Tremolite, named after Tremola, a valley in the Alps, where it was discovered, is a white variety of amphibole, or hornblende, occurring in long, bladelike crystals, and coarsely fibrous masses. The fibrous variety has been used for asbestos. The compact variety is called nephrite and is used for ornamental purposes in the orient. It is most often found in impure limestones where the rock has re-crystallized during metamorphism.

Trenton Beds

Trenton beds are beds of rocks of the Trenton Period, so named from the typical example of the beds found at Trenton Falls, New York, USA.

Trenton Period

In geology, the Trenton period is a subdivision in the lower Silurian system of rocks It is so named from Trenton Falls, in New York, USA, a typical example of the system. The rocks are mostly limestones and black carbonaceous shales, and the period is divided into the Trenton, Utica, and Cincinnati epochs. The rocks occur in many localities in north America and are chiefly used in building.

Trias

The trias is the formation situated between the Permian and Lias, and so named by the Germans, because it consists of three series of strata, which are called in German the Bunter Sandstein, Muschelkalk, and Keuper.

Triassic

In geology, the Triassic (named after the Trias) was the ninth geological period, 170,000,000 years ago, following the Carboniferous and preceding the Jurassic periods. The rocks are the oldest of the Mesozoic formations and, once known as the New Red Sandstone are now divided into three series: the Keuper, Muschelkalk and Bunter. The rocks of the Triassic period are chiefly red or mottled sandstones or shales and limestones, and red marls, and often reach great thickness.

The rocks of the period are widespread, though in America they grade insensibly into the Jurassic and are often designated Jura-Trias. In the Midlands of England and in Scotland and Ireland, Triassic rocks cover large areas, those at Northwich, Middlewich, Droitwich etc being important sources of rock salt. Triassic rocks cover practically all of

Germany and large areas in Europe, north and south America, Africa, Asia and Australasia.

Trichroism

Trichroism is the quality possessed by some crystals of presenting different colours in three different directions.

Triclinic

Triclinic refers to a crystal with six faces as parallelograms and three axes of unequal length all inclined to each other. An uncommon form of crystal.

Tridymite

Tridymite is a form of pure silica, like quartz, but having a lower specific gravity and crystallizing in hexagonal tables. It is found in trachyte and similar rocks, such as the acidic volcanic rocks at San Cristobal in Mexico, and is also found in the Euganean Hills and at Siebengebirge on the Rhine.

Tripestone

Tripestone is a variety of anhydrite composed of contorted plates thought to resemble pieces of tripe.

Triphylite

Triphylite is a mineral of a greyish-green or bluish colour, consisting of the phosphates of iron, manganese, and lithia. A salmon-coloured or clove- brown variety containing very little iron is known as lithiophilite.

Triplite

Triplite is a mineral of a dark brown colour, generally with a fibrous, massive structure. It is a fluophosphate of iron and manganese.

Tuff

Tuff is another name for tufa.

Tungstite

Tungstite is the oxide of tungsten, a yellow mineral occurring in a pulverulent form. It is often associated with wolfram.

Turnerite

Turnerite, named after the English chemist and mineralogist C. H. Turner, is a variety of monazite.

Turonian

Turonian is one of the geological subdivisions into which the Upper Cretaceous formation of Europe is divided.

Turquoise

Turquoise (also known as calaite) is a mineral of secondary origin usually found in small veins and stringers. Turquoise is a hydrous phosphate of alumina containing a little copper. It has a blue, or bluish green, colour, and usually occurs in reniform masses with a botryoidal surface. It has a relative hardness of 6. Turquoise is susceptible of a high polish, and when of a bright blue colour is much esteemed as a gem. The finest specimens come from Persia. It is also found in New Mexico and Arizona, and is regarded as identical with the chalchihuitl of the Mexicans.

Twin

Twin refers to a mineral specimen comprised of two or more single crystals intergrown in a systematic arrangement.

Typolite

A typolite is a stone or fossil which has on it impressions or figures of plants and animals.

Tyrolite

Tyrolite, named after the Tyrol, where it occurs, is a translucent mineral of a green colour and pearly or vitreous lustre. It is a hydrous arsenate of copper.

Tysonite

Tysonite, named after S.T. Tyson, is a fluoride of the cerium metals occurring in hexagonal crystals of a pale yellow colour.

Ulexite

Ulexite or hayesine is an evaporate mineral and an ore of boron. It is formed during the evaporation of lake basins and occurs in white rounded crystalline masses. It is a hydrous borate of lime and soda and has a relative hardness of 2. Ulexite was named in honour of George Ulex, the German chemist who in 1850 first correctly analysed the mineral. One transparent variety of ulexite exhibits the rare property of transmitting light unaltered from one end of its crystals to the other. Ulexite is utilized as a source of boracic acid, and for glass and pottery manufacture.

Ullmannite

Ullmannite, named after J.C. Ullman, the German chemist, is a brittle mineral of a steel-grey colour and metallic lustre, containing antimony, arsenic, sulphur, and nickel.

Umber

Umber is a naturally occurring mineral used as a pigment.

Underclay

Underclay is a stratum of clay lying beneath a coal bed, often containing the roots of coal plants, especially the Stigmaria.

Uralite

Uralite, so called because it was first observed in the Ural Mountains, is an amphibole resulting from the alternation of pyroxene by paramorphism. It is not uncommon in massive eruptive rocks. Uralite is green in colour.

Uralitization

Uralitization is the change of pyroxene to amphibole by paramorphism.

Uraninite

Uraninite is a radioactive important source of uranium. It consists chiefly of uranium oxide with some lead, thorium, etc. It occurs in pegmatites and veins in granite; also as sedimentary deposits. It is found in black octahedrons and also in masses with a pitch like lustre. It has a relative hardness of 6.

Utica

Utica is a subdivision of the Trenton Period of the Lower Silurian, characterized in the State of New York by beds of shale.

Vanadinite

Vanadinite is a source of vanadium and a minor ore of lead (consisting of lead vanadate with a small proportion of lead chloride). It is found occurring in yellowish, and ruby-red hexagonal crystals in the oxidized portion of lead veins associated with other lead minerals. Vanadium is used as a steel- hardening metal. It has a relative hardness of 3.

Variolite

Variolite is a kind of diorite or diabase containing imbedded whitish spherules, which give the rock a spotted appearance.

Variscite

Variscite, named after Variscia in Germany, is an apple-green or pink coloured secondary mineral occurring in reniform masses. In its pure form it is a hydrated phosphate of aluminium. Variscite was confirmed as a distinct species of mineral in 1837.

Vauquelinite

Vauquelinite, named after the French chemist Vauquelin, is a chromate of copper and lead, of various shades of green.

Vein

In geology a vein is a narrow mass of rock intersecting other rocks, and filling inclined or vertical fissures not corresponding with the stratification. Such mineral ores have usually been deposited from solutions, and may be metallic or non-metallic; as a general rule, the two kinds are found in veins together. The non-valuable mineral deposit in a vein is known as the gangue. Veins are sources of many of the valuable metals, e.g. gold and silver, and vary in size from the almost invisible to several hundred feet in thickness.

Venus' Hair

Venus' Hair is a popular name for a variety of quartz penetrated by acicular crystals of rutile.

Vermiculite

Vermiculite is a group of minerals having, a micaceous structure. They are hydrous silicates, derived generally from the alteration of some kind of mica. They are so called because the scales, when heated, open out into worm- like forms. Processed vermiculite is used as an insulating material in building.

Vesicle

Vesicle refers to a small cavity in a volcanic rock.

Vesuvianite

Vesuvianite (also known as idocrase) is a mineral usually found in crystalline limestones occurring in tetragonal crystals, and also massive, of a brown to green colour, rarely sulphur yellow and blue, and is formed as a result of contact metamorphism. It is a silicate of alumina and lime with some iron magnesia and has a relative hardness of 7. Vesuvianite was confirmed as a distinct species of mineral in 1795 and named vesuvianite after Mount Vesuvius in Italy where typical specimens are found.

Vicinal Planes

Vicinal planes are subordinate planes on a crystal, which are very near to the fundamental planes in angles, and sometimes take their place. They have in general very complex symbols.

Viridite

Viridite is a greenish chloritic mineral common in certain igneous rocks, such as diabase, as a result of alternation.

Vitreous

Vitreous refers to a lustre like that of glass. Quartz is an example.

Vivianite

Vivianite, named after the English mineralogist J G Vivian who discovered the mineral in Cornwall, England, is a rare mineral of secondary origin, associated with pyrite in copper and tin veins. It forms as a weathering product from primary iron- manganese phosphates in pegmatites, and is a hydrous phosphate of iron of a blue to green colour, growing darker on exposure to the air. It occurs in monoclinic crystals, also fibrous, massive, and earthy. It is also found in beds of clay and has a relative hardness of 1.5 to 2. Vivianite was confirmed as a distinct species of mineral in 1817.

Volborthite

Volborthite, named after its discoverer, Volborth, Is a mineral occurring in small six-sided tabular crystals of a green or yellow colour. It is a hydrous vanadate of copper and lime.

Voltzite

Voltzite (voltzine), named in honour of the French engineer Voltz, is an oxysulphide of lead occurring in implanted spherical globules of a yellowish or brownish colour.

Vulpinite

Vulpinite, named after Vulpino, in Italy, is a scaly granular variety of anhydrite of a greyish white colour, used for ornamental purposes.

Wacke

Wacke is a soft, earthy, dark-coloured rock or clay derived from the alteration of basalt.

Wagnerite

Wagnerite is a fluophosphate of magnesia, occurring in yellowish crystals, and also in massive forms.

Warwickite

Warwickite is a dark brown or black mineral, occurring in prismatic crystals imbedded in limestone near Warwick, New York. It consists of the borate and titanate of magnesia and iron.

Water Sapphire

Water sapphire is a deep blue variety of iolite, sometimes used as a gem.

Wavellite

Wavellite, named after its discoverer William Wavel who found it in clay near Barnstaple, Devon, England, is a rare mineral of secondary origin confirmed as a distinct species of mineral in 1805. It is a hydrous phosphate of alumina, occurring usually in hemispherical radiated forms varying in colour from white to yellow, green, or black. It is found frequently in small amounts in crevices in aluminous, low-grade metamorphic rocks and in limonite and phosphorite deposits. It has a relative hardness of 4.

Wealden

Wealden is the lowest division of the Cretaceous formation in England and on the Continent, which overlies the Oolitic series.

Websterite

Websterite (also known as aluminite, hallite) is a white or grey nodular hydrous mineral named after the geologist Webster, is a hydrous sulphate of alumina occurring in white reniform masses. Websterite forms monoclinic crystals, generally too small to see with the naked eye.

Wenlock Group

The Wenlock group is the middle subdivision of the Upper Silurian in Great Britain. It is so named from the typical locality in Shropshire.

Wernerite

Wernerite is the common greyish or white coloured variety of soapolite.

Wheel Ore

Wheel ore is a variety of bournonite so named from the shape of its twin crystals.

Whet Slate

Whet slate (whetstone slate or oilstone) is a variety of slate used for sharpening cutting instruments.

Whewellite

Whewellite, named after Professor William Whewell of Cambridge, England, is calcium oxalate, occurring in colourless or white monoclinic crystals.

Whitneyite

Whitneyite, after the American geologist J.D. Whitney, is an arsenide of copper from Lake Superior.

Willemite

Willemite (troostite), named after Willem I, king of the Netherlands, is a mineral first confirmed as a distinct species of mineral in 1830. It is an ore of zinc found in crystallized limestone and associated with calcite, zincite, franklentie. It is a silicate of zinc, usually occurring massive and of a greenish yellow colour, also in reddish crystals containing manganese. It has a relative hardness of 6.

Withamite

Withamite, named after its discoverer, H. Witham, is a variety of epidote, of a reddish colour, found in Scotland.

Witherite

Witherite, named after Dr. Wiliam Withering, is a somewhat rare mineral found most often in veins associated with galena. Witherite was confirmed as a distinct mineral in 1789 and is a barium carbonate occurring in white or grey six-sided twin crystals, and also in columnar or granular masses and is a minor source of barium.

Wolframite

Wolframite is tungstate of iron and manganese, generally of a brownish or greyish black colour, sub-metallic lustre, and high specific gravity. It occurs in cleavable masses, and also crystallized. It is the most important tungsten ore and has a relative hardness of 5.

Wollastonite

Wollastonite, named after the English chemist William Wollaston, is a silicate of lime of a white to grey, red, or yellow colour, occurring generally in cleavable masses, and rarely in tabular crystals, mainly as a contact metamorphic mineral in crystalline limestones.

Wood Opal

Wood opal is a striped variety of coarse opal, having some resemblance to wood.

Wulfenite

Wulfenite (Yellow lead ore), named after F. X. Wulfen, an Australian mineralogist, is native lead molybdate occurring in tetragonal crystals, usually tabular, and of a bright orange-yellow to red, grey, or brown colour. It has the formulae $PbMoO_4$ and a relative hardness of 3. It is a minor source of molybdenum and is found in the oxidized portion of lead veins with other secondary lead minerals.

Wurtzite

Wurtzite is a rare mineral formed mainly with other sulphide ores in hydrothermal veins. It is sometimes also found in concretions in sedimentary rocks such as clay and ironstone and in contact metamorphosed limestone.

Wurtzite has an unusual crystal symmetry which can be either hexagonal or trigonal depending upon how the specimen formed. Wurtzite is a minor ore of zinc and was confirmed as a distinct species of mineral in 1861 and named after Karl Wurtz, the French chemist.

Xenotime

Xenotime is a native phosphate of yttrium occurring in yellowish-brown tetragonal crystals.

Yellow Copperas

Yellow copperas (copiapite) is a yellow coloured secondary mineral of the copiapite group. A hydrous sulphate of iron it is formed through the oxidation of iron sulphide deposits and is associated with pyrite. Externally the crystals appear in tabular shapes, in plate-like aggregates of micro crystals, or in powdery encrustations on the surface of rocks and other minerals.

Yellow Ochre

Yellow ochre is a naturally occurring mix of limonite and clay.

Yttro-cerite

Yttro-cerite is a mineral of a violet-blue colour, inclining to grey and white. It is a hydrous fluoride of cerium, yttrium, and calcium.

Yttro-Columbite

Yttro-Columbite is a tantalate of uranium, yttrium, and calcium, of a brown or black colour.

Zaratite

Zaratite (emerald nickel), named after General Zarata of Spain, is a hydrous carbonate of nickel occurring as an emerald-green incrustation on chromite.

Zarnich

Zarnich is native sulphide of arsenic, including sandarach, or realgar, and orpiment.

Zechstein

Zechstein is the upper division of the Permian (Dyas) of Europe. The prevailing rock is a magnesian limestone.

Zeolites

Zeolites are a group of complex silicate minerals mostly occurring in cavities in basic volcanic rocks. They are usually colourless or white and are relatively soft. Formerly used as water softeners, they are used as molecular sieves in the petroleum industry and as drying agents.

Zietrisikite

Zietrisikite is a mineral wax, very similar to ozocerite. It is found at Zietrisika, Moldavia, hence its name.

Zincite

Zincite is native zinc oxide. It is a brittle, translucent mineral, of an orange-red colour and has the formulae (Zn,Mn)O and a relative hardness of 5. It is an ore of zinc and used in the production of zinc oxide.

Zinkenite

Zinkenite, named after Zinken, director at one time of the Hanoverian mines, is a steel-grey coloured metallic mineral, a sulphide of antimony and lead.

Zinnwaldite

Zinnwaldite, named after Zinnwald, in Bohemia, where it occurs, is a kind of mica containing lithium, often associated with tin ore.

Zircon

Zircon (zirconium silicate) has the formulae $ZrSiO_4$ and a relative hardness of 8. It is a common accessory mineral found in all types of igneous rocks. When transparent it's used as a gemstone.

Zircons contain abundant radioactive isotopes of uranium and so are useful for uranium-lead dating to determine the ages of rocks.

Zoisite

Zoisite, named after its discoverer its discoverer, Von Zois, an Austrian mineralogist, is a greyish or whitish mineral allied to epidote, occurring in orthorhombic, prismatic crystals, also in columnar masses and formed in high temperature metamorphic rocks. It is a silicate of alumina and lime has the formulae Ca_2Al_3 $(Si_3O_{12})(OH)$ and a relative hardness of 7.

Zunyite

Zunyite is a fluosilicate of alumina occurring in tetrahedral crystals in Colorado.

Index

N

O

❑❑❑